ON TIME AND WATER

Copyright © Andri Snær Magnason, 2019
Title of the original Icelandic edition: **Um tímann og vatnið**
Published by agreement with Forlagið, www.forlagid.is
Translation copyright © 2021 by Lytton Smith

First edition, 2021
First Open Letter paperback edition, 2022
All rights reserved

Library of Congress Cataloging-in-Publication Data: Available.
ISBN-13: 978-1-948830-53-9

This project is supported in part by an award from the National Endowment
for the Arts and the New York State Council on the Arts with the support of the
Governor of New York and the New York State Legislature.

This book has been translated with financial support from

 ICELANDIC LITERATURE CENTER

Cover designed by Zoe Norvell

Open Letter is the University of Rochester's nonprofit, literary translation press:
Dewey Hall 1-219, Box 278968, Rochester, NY 14627

www.openletterbooks.org

Printed in the United States

ON TIME AND WATER

Andri Snær Magnason

translated by LYTTON SMITH

OPEN LETTER
LITERARY TRANSLATIONS FROM THE UNIVERSITY OF ROCHESTER

*"In the end, we will conserve only what we love,
we will love only what we understand,
and we will understand only what we are taught."*

Baba Dioum

Contents

May you live in interesting times

"Take notice what you notice."

—Thorvaldur Thorsteinsson

Whenever I host overseas visitors to Reykjavík, I like to drive them along Borgartún, a street I call the Boulevard of Broken Dreams. I point out Höfdi, the white wooden house where Ronald Reagan and Mikhail Gorbachev met in 1986, a house that many people associate with the end of communism, the fall of the Iron Curtain. The nearest building to Höfdi is a black boxy structure, all glass and marble, that once housed the headquarters of Kaupthing Bank. Kaupthing's collapse in 2008 was the fourth-largest bankruptcy in the history of capitalism—not merely per capita of the Icelandic population but in net U.S. dollars: 20 billion dollars.[1]

I don't mean to gloat over others' misfortunes, but it astonishes me that before middle age I'd already witnessed the collapse of two vast belief systems, communism and capitalism. Each had been maintained by people who'd scaled the peaks of the establishment, of government and of culture, people esteemed in direct proportion to their relative position at the pyramid's apex. Deep inside these systems, people kept up appearances right to the bitter end. On January 19, 1989, the East German General Secretary of the Socialist

Unity Party of Germany, Erich Honecker, said: "The wall will stand in fifty years' time, and a hundred years', too." The wall collapsed that November. Kaupthing's CEO said in a television interview on October 6, 2008, after the bank had received emergency loans from the Central Bank of Iceland: "We're doing very well indeed, and the Central Bank can be confident it will get its money back . . . I can tell you that without hesitation." Three days later, Kaupthing collapsed.

When a system collapses, language is released from its moorings. Words meant to encapsulate reality hang empty in the air, no longer applicable to anything. Textbooks are rendered obsolete overnight and overly complex hierarchies fade away. People suddenly find it difficult to hit upon the right phrasing, to articulate concepts that match their reality.

Between Höfdi and Kaupthing's former headquarters there's a grassy lawn. In its center stands a paltry copse of trees: six spruces and some woolly willow shrubs. Lying inside that cluster of trees, between the two buildings, looking up at the sky, I found myself wondering which system would collapse next, what big idea would be the next to take hold.

Scientists have shown us that the foundations of life, of Earth itself, are failing. The principal ideologies of the twentieth century considered the Earth and nature as sources of inexpensive, infinite raw material. Humans assumed that the atmosphere could continually absorb emissions, that oceans could endlessly absorb waste, that soil could constantly renew itself if given more fertilizer, that animal species would keep moving aside as humans colonized more and more space.

If scientists' predictions prove accurate about the future of the oceans and the atmosphere, about the future of weather systems, about the future of glaciers and coastal ecosystems, then we must ask what words can encapsulate these immense issues. What ideology can handle this? What should I read? Milton Friedman, Confucius, Karl Marx, the Book of Revelation, the Koran, the Vedas? How to tame these desires of ours, this consumption and materialism that, by any and every measurement, promise to overpower Earth's fundamental life systems?

This book is about time and water. Over the next hundred years, there will be foundational changes in the nature of water on our Earth. Glaciers will melt away. Ocean levels will rise. Increasing global temperatures will lead to droughts and floods. The oceans will acidify to a degree not seen for fifty million years. All this will happen during the lifetime of a child who is born today and lives to be my grandmother's age, ninety-five.

Earth's mightiest forces have forsaken geological time and now change on a human scale. Changes that previously took a hundred thousand years now happen in one hundred. Such speed is mythological; it affects all life on Earth, affects the roots of everything we think, choose, produce, and believe. It affects everyone we know, everyone we love. We are confronted by changes that are more complex than most of what our minds typically deal with. These changes surpass any of our previous experiences, surpass most of the language and metaphors we use to navigate our reality.

Compare it to trying to record the sounds of a volcanic eruption. With most devices, the sound becomes muddled; nothing can be heard but white noise. For most people, the phrase "climate change" is just white noise. Easier to have opinions on smaller matters. We can comprehend the loss of something valuable, can comprehend when an animal is shot, when a project blows past its agreed-upon budget. But when it comes to the infinitely large, the sacred, to things that are fundamental to our lives, there's no comparable reaction. It's as if the brain cannot register at such a scale.

This white noise deceives us. We see headlines and think we understand the words in them: "glacial melt," "record heat," "ocean acidification," "increasing emissions." If the scientists are right, these words indicate events more serious than anything that has happened in human history up to now. If we fully understood such words, they'd directly alter our actions and choices. But it seems that 99 percent of the words' meanings disappear into white noise.

Perhaps "white noise" is the wrong metaphor; the phenomenon is more like a black hole. No scientist has ever seen a black hole, which can have the mass of millions of suns and can completely absorb light. The way to detect black holes is to look past them, to look at nearby nebulae and stars. When it comes to discussing issues that affect all water on Earth, all Earth's surface, the planet's entire atmosphere—the issue's enormity absorbs all the meaning. The only way to write about the subject is to go past it, to the side, below it, into the past and the future, to be personal and also scientific and to use mythological language. I need to write about things by *not* writing about them. I need to go backward to move forward.

We're living in a time when thought and language have been freed from ideological chains. We're living in the time of that old Chinese curse, almost surely translated incorrectly, yet no less apt for that: "May you live in interesting times."

A little treasure

I graduated with a literature degree from the University of Iceland in 1997. That summer, I worked in the basement of the Árni Magnússon Institute for Medieval Studies. The Institute was behind a locked door on the university campus; for some reason, despite having studied in this same building for years, I'd never once gone through that particular door. It was a mysterious threshold, like the so-called elvish rocks where Iceland's "hidden people" are said to live. I'd heard stories of people who'd disappeared through the door never to reemerge. Inside, manuscripts of Icelandic sagas were housed alongside scholars who needed peace and time to pore over these treasures. Ringing the doorbell was so daunting that it felt more like pulling a fire alarm. I didn't dare push the button until the day I found myself filled with a burning desire to see what lay inside. I rang and was invited in.

Behind the door it was quiet and twilit; the air hung with a heavy smell of old books, a stillness that was truly overwhelming for a young man. I felt a pang of unease. I was here, inside, among the scholars of old manuscripts, some of them my grandparents' contemporaries. I was struck by my own insignificance when discussions around the carafe turned to whether Thorvald had been in Skagafjördur in the summer of '86: I had no idea whether they were talking about 1186, 1586, or 1986. My fear that I'd be considered poorly read increased in tandem with my inarticulacy. I felt myself simultaneously (did I mean *concurrently*?) witless and ungrammatical.

I'd always done outdoor jobs in the summers—paving and gardening. I'd pitied office workers for having such little freedom. Inside the Institute, I'd catch myself looking out the window at my lightly dressed peers mowing the grass on the university green while my mind traveled farther out, past them, into the wider world. John Thorbjarnarson, my mom's half-brother, a biologist, had invited me to come and help him study the mating habits of anacondas in the Venezuelan mangroves. We'd also work with a team of scientists in the Amazon rainforest, counting crocodile eggs in the Mamirauá Sustainable Development Reserve as part of efforts to save the black caiman crocodile, *Melanosuchus niger*, the largest predator in South America.[2] The water level in the flooded forest fluctuates about ten meters a year, so we would be staying in floating houses. "It's no small joy to wake up in the morning to the sound of dolphins fishing right outside your door," John remarked.

At the time, my girlfriend Margrét and I were expecting our first child; it would have been somewhat irresponsible of me to charge off on such an adventure. You might say two tracks diverged in my life. The train steamed off to Venezuela and on to the Amazon without me, leaving me a sort of bystander to my own existence, doubtful whether serious scholarship and the solitude of writing suited me.

One day, I was asked to docent a manuscript exhibition in a small gallery on the upper floor. The philologist Gísli Sigurdsson was in charge of the exhibition; he told me to follow him to a hefty steel door in the basement, where he produced three keys. When he opened the door, I was amazed to see this was where the manuscripts lived: this was the sacred heart of Icelandic cultural history. I was surrounded by awe-inspiring historic gems. Inside were vellum manuscripts, the oldest of which had been written around 1100, depicting events from bygone ages. Inside were the original manuscripts of the Icelandic sagas: about Vikings and knights, about kings, ancient lawbooks. Gísli went to a shelf and opened a box. He brought out a small manuscript, and gently handed it to me.

"What's this book?" I whispered.

I don't know why I was whispering. It simply felt right to use a hushed tone in that space.

"This is the *Codex Regius. Konungsbók,* the *King's Book.*"

I felt weak at the knees, starstruck. *Codex Regius,* containing the Poetic Edda, is the greatest treasure in all of Iceland, perhaps even the whole of Northern Europe: the second major source of Nordic mythology and the earliest manuscript of the famous poems "Völuspá," "Hávamál," and "Thrymskvida." A primary inspiration for Wagner, Borges, and Tolkien. I felt like I had Elvis Presley himself in my arms.

The manuscript was unassuming. Considering its contents and its influence, it ought to have been shining golden, resplendent. In reality, it was small and dark, almost like a book of spells. It was timeworn but not wizened, a beautiful brown vellum with simple, distinct lettering and almost no illustration beyond a few initial capitals. It offered the oldest evidence that a book shouldn't be judged by its cover.

The philologist carefully opened the manuscript and showed me a clearly legible *S* in the middle of the page. "Read that," he said, so I squinted at the script until I could read it myself: *Sól tér sortna sígur fold í mar . . .* (The sun goes dark, the lands sink, the shining stars disappear from the skies, the great ash will burn and fierce heat will lick the skies).

A shiver ran down my spine: this was Ragnarök itself, the end of the world as described in the original prophecy in the poem "Völuspá." The sentences were all in one continuous line, not broken up into verses the way poems usually are when printed in books. I was in direct contact with whomever had written these words down on the page over seven hundred years ago. I became hypersensitive to my environment, scared to cough, fearing I would drop it, feeling a touch guilty for even breathing this close to the book. Perhaps that was an overreaction; after all, this manuscript had been stored in damp turf houses for five centuries and transported on horseback in chests across surging glacial rivers; in 1662, it had been sent by ship to Denmark as a gift to King Frederick III. I felt an extreme sensation, a connection to deep time. I spoke practically the same language as the person who'd written the manuscript. Could it last another seven hundred years? Until the year 2700? Would our language and civilization live that long?

As a species, humankind has preserved relatively few of its sacred ancient mythologies: ideas about the forces and gods that ruled the heavens and about the genesis and end of the world. We have Greek, Roman, Egyptian, and Buddhist mythologies. We have the Hindu worldview, that of Judeo-Christians and of Islam, fragments of the Aztec worldview. Nordic mythology is one such worldview; for that reason, the *Codex Regius* is more important than even the *Mona Lisa*. Most of what we know about the Nordic gods, Valhalla, and Ragnarök, comes from this book. The manuscript is a never-ending inspiration, a wellspring for beliefs and for art. From it come works of modern dance, death metal bands, even contemporary Hollywood classics such as Marvel Studios' *Thor: Ragnarok*, in which Thor and his friend Hulk combine their might against the treacherous Loki, the giant Surtur, Hela, and the terrifying Fenris Wolf.

I put the manuscript in a small dumbwaiter and sent it up to the next floor. In the meantime, I galloped up a flight of narrow stairs to meet it. I set it with extreme care on a little cart that I pushed down a long corridor. I locked it securely inside a glass case where it was protected like a premature baby in an incubator. That whole week my dreams were troubled. In them, I was usually downtown and had managed to lose the book. One time, I met a woman with a cleaning cart walking along the corridor; I foresaw a cultural disaster, the manuscript falling into a pail of soapy water and bobbing up squeaky clean, pages blank. A tabula rasa.

Marketing was not a strong suit of the medieval scholars at the Institute, so I spent whole days alone with these treasures while tourists shuttled out to waterfalls and geysers, to Gullfoss and Geysir. It was certainly privilege enough to hang out with this *Mona Lisa* of ours, but there was so much more besides. The Institute's foremost gems were also on display alongside the *King's Book*: *Grágás*, containing the Viking Age laws; *Mödruvallabók*, containing the major Icelandic sagas; *Flateyjarbók,* with its two hundred calfskin sheets and vivid illustrations. I sometimes stood over the glass boxes trying to read the text on the open pages. The *King's Book* was the most legible, its lettering clear enough I could stumble through these ancient words: *Ungur var eg fordum, fór eg einn saman* . . . (When I was young and set out on my own, I lost my way. I thought myself rich when I came into company. Man is a joy to man).

This was the same week Margrét and I rushed to the maternity ward in the middle of the night and I held my newborn son in my arms. Never had I handled anything so novel and so delicate. Nor had I handled anything so old and so delicate. I continued dreaming I was downtown, but now suddenly aware I was only in my underwear, and had lost both my son *and* the manuscript.

In the room next to the manuscript vault were more treasures: a hoard of tapes, recordings folklorists had collected all around Iceland between 1903 and 1973. There one could hear the oldest recording in Iceland, set on wax cylinder, Edison's "Graphophone," in 1903. There were old women, farmers, and sailors singing lullabies and chanting ancient rhymes, telling stories. I'd never heard anything so strange and beautiful and the thought flew into my head that it was urgent these ancestral voices reach the ears of the general public. My chief task that summer was to work with the folklorist Rósa Thorsteinsdóttir, compiling a selection from the archive so we could put out a CD.

Each time I threaded the black coils on the tape player and put on the headphones, I stepped into a time machine. I was in the living room of an old woman who had been born in 1888. The clock was ticking in the kitchen and she was reciting a rhyme she had learned from her grandmother, who had been born in 1830, a rhyme she had learned from her grandmother, who was born in the late eighteenth century during the time of the great Laki volcanic eruption—a rhyme she had learned from her grandmother, who was born in 1740. The recording was made in 1969, so the cycle spanned nearly 250 years. It was from a world in which the eldest taught the youngest. The old-fashioned aesthetics of the rhymes differed from our idea of beautiful singing. The vocal tone and singing style did not resemble anything I'd ever come across. I collected samples on a reel and let my friends listen and asked them to guess where the music was from. They guessed Native Americans, Sami reindeer herders, Tibetan monks, Arabic prayers. When they had listed all the remotest cultures they could think of, I said, "This was recorded here, in the Westfjords, in 1970. The man you hear singing was born in 1900."

I played the recordings at home for my son when he was fussy; he calmed down as soon as the melodies started up. (I had half a mind to do a scientific study into whether ancient chanting has a marked sedative effect on infants.)

I was fascinated by the idea of capturing time. I realized how much around me would soon disappear like the women on those slick black reels. I had five living grandparents and that summer I began collecting their stories in haphazard fashion: Grandpa Jón was born in 1919, Grandma Dísa in 1925, Grandma Hulda in 1924, Grandpa Árni (Hulda's second husband) in 1922, and Grandpa Björn in 1921. Theirs was a generation at an unparalleled turning point, born just after World War I and living through the era of the Great

Depression. They lived through World War II as well as many of the greatest changes of the twentieth century. Some of them were born before the time of electrical lighting and machines, born into a society of poverty and hunger. Inspired by the reel collection, I decided to interview people close to me. I used a handy VHS recorder, a Dictaphone, and then my smartphone once that technology arrived. I really didn't know what I was looking for, I was just trying to collect anything I could so that people in the future might appreciate it. I was making my own archive: the Andri Magnason Institute.

A future conversation

I'm at Grandma Hulda and Grandpa Árni's home in Hladbær. We're sitting in the kitchen: the Ellidaá stream meanders in front of the house and people jog along the river path. A few snowdrifts linger on the slopes of Bláfjöll yet the garden is in full bloom. I open my computer and load a video so I can show Grandma Hulda and my mother a film no one has seen in decades. I'd discovered an old 16 mm cassette in their storage room and had converted it to digital. It was a movie Grandpa Árni shot in 1956, black-and-white and silent; the picture quality is perfect. Well-mannered children sit in the dining room here at number 3 Selás, this big white house my great-grandpa built on the banks of the stream. The children have little bottles of cola; Grandma Hulda appears, smiling, with a magnificent pavlova decorated with lit candles. At the end of the table, ten-year-old twin sisters sit together, laughing and blowing vigorously at the candles. Great-Grandma is there, too, dressed in traditional Icelandic fashion, watching it all. The next shot shows the children dancing in a ring in the yard; no doubt they're playing the game "In a Green Hollow." Mom and Grandma Hulda watch the video and name the people the images have preserved. A child's birthday from 1956 captured on 16 mm film is truly something. There isn't even footage of the Icelandic government from that time.

And here we are in 2018, sitting in the same kitchen more than sixty years later. Mom is over seventy, Grandma Hulda is ninety-four years old, and my youngest daughter is ten. Grandma Hulda has hardly changed from how I remember her: she's only just given up golfing and her memory is still intact. A few years ago, a man who was trying to sing her praises to me commended how *sharp* she was. I acted half offended: *sharp*? What do you mean, *sharp*; she's always been a quick thinker. She certainly doesn't think of herself as elderly. Take her sense of humor. That's a beautiful shawl, I once remarked about a blue shawl she was wearing. Yes, an old lady crocheted it for me. Old lady? I asked. She laughed and replied: Oh, yes, she's probably ten years younger than I am!

The phone rings and Grandma Hulda runs to answer it. We sit down to eat pancakes as the radio hums low in the background. I ask Hulda Filippía, my daughter, to do a little math puzzle.

"How old is your great-grandma if she was born in 1924?"

"She's ninety-four," Hulda replies immediately.

"Fast math," I say.

"Well, I know how old she is." She grins.

"All right, but now you'll really have to calculate. When will *you* be ninety-four?"

"So it would be the year I was born, 2008, plus ninety-four?"

"Exactly."

She takes a piece of paper and a pen and looks skeptically at the sheet. She shows me the result as though it must be a misunderstanding.

"Is that really right, 2102?"

"Yes, hopefully you'll be just as energetic as Grandma Hulda is now. Maybe you'll even be living in this same house. Maybe your ten-year-old great-granddaughter will be visiting, sitting with you in this kitchen in 2102, just like you are sitting here right now."

"Yes, maybe," says Hulda, sipping a glass of milk.

"One more equation. When will your great-granddaughter be ninety-four years old?"

Hulda writes some figures on a piece of paper, with a little help.

"Would she have been born in 2092?"

"Yes, that's right."

"Okay, 2092 plus ninety-four . . . 2186!

She laughs at the thought.

"Yes, can you imagine that? You, born in 2008, might know a girl who will still be alive in the year 2186."

Hulda purses her mouth and looks into the air.

"Can I go now?" she asks.

"Almost," I say. "I've one more puzzle. How long is it from 1924 to 2186?"

Hulda does the math.

"Is it two hundred and sixty-two years?"

"Imagine that. Two hundred and sixty-two years. That's the length of time you connect across. You'll know the people who span this time. Your time is the time of the people you know and love, the time that molds you. And your time is also the time of the people you *will* know and love. The time that you will shape. You can touch two hundred and sixty-two years with your bare hands. Your grandma taught you, you will teach your great-granddaughter. You can have a direct impact on the future, right up to the year 2186."

"Up to 2186!"

A projection

In the TV room, Grandpa Árni has pulled down the roller blind and set up his slide projector on an old ironing board. He fetches a tray of photographs from his little office. They're lined up in rows with his handwriting on the front of each one: *Lónsöræfi 1965*, *Vatnajökull 1955*, *Kverkfjöll 1960*. There are snowmobiles, ski cabins, ski champions.

An image appears on the screen, a man shoveling snow that is at least six meters deep; the prow and engine of a large plane are visible, jutting out from the drift. Grandpa Árni can remember almost everything from a long time ago, especially if the memory is accompanied by a photograph.

"This picture was taken at Bárdarbunga on Vatnajökull glacier in 1951. There we are shoveling out a U.S. Army Skytrain plane. It had flown to rescue the crew of another stricken plane, the *Geysir*, but wasn't able to take off again, so it was left behind on the glacier. That winter, a thick layer of snow buried the plane. Several enterprising local pilots contacted the military; they bought the aircraft for seven cents per pound, then they launched an expedition to retrieve it.

"Back then, we'd only just started the FBSR—Reykjavík's Air Ground Rescue Team," says Grandpa Árni. "We located the prop plane—she was like a little bump on the glacier—and dug it out. It was hellish work: the thing was inside a seven-meter-high snowdrift. We dragged her down to a temporary runway, and it sparked to life first try. From there, it was flown to Reykjavík. It was something of a fairy tale," Grandpa Árni says wistfully.

The wreck of the *Geysir* lay nearby and they had been able to save some of the goods that were still on the plane.

Grandma Hulda jumps into the conversation.

"Gudbjörg's christening dress was made from cloth that had sat all winter up on the glacier, in the plane."

Grandpa Árni shows me pictures of the wreck; in them I can see the nose wheel and the aircraft's name. *Geysir* had lost its way and crashed into the middle of Vatnajökull. The crew was presumed dead, the search called off, and a period of national mourning observed while memorials were planned. By chance, a coast guard ship sailing off the coast off Langanes picked up an emergency call, ". . . CIER." No one could understand it until they realized it was the second half of the word "glacier."

The plane was found after an extensive search. The U.S. Army rescue plane flew to the scene a week after the crash and landed on the glacier, but the aircraft couldn't take off again as its skis had frozen to the ice. Fortunately, rescuers came from the northern city of Akureyri, and with their help everyone came down off the glacier, including the six-person crew of the *Geysir*, either stumbling along or on skis. It's only a matter of time before someone makes a film about this plane wreck.

The albums contain old films from a Rolleiflex camera; there are boxes of 8 mm and 16 mm film from the Bolex camera Grandpa Árni bought off the artist Gudmundur Einarsson. There are thousands of photographs and films, spanning almost his entire life. He is rarely in any of the pictures; usually he's behind the lens. My mom is there, or my moms, as I call them to tease them, Kristín and Gudrún, the then eleven-year-old identical twin sisters in overalls with *KB* and *GB* stitched in big letters on the pants to distinguish them.

These are priceless resources; some of the films and images are absolute works of art. Grandpa Árni clearly had an eye for beautiful framing. He bought himself a scanner and a computer when he was eighty; most people thought he was delusional, that some rogue salesman had tricked him. But over the next few years, he spent entire days in his little office scanning all the old film. At first, he printed out the photos, but after he turned ninety, he started posting them right to Facebook. There's material there for novels, books of photography, documentaries. So many stories that could be told.

My son is going through a photography phase; Grandpa Árni recently gave him his old Rolleiflex camera from 1960. He even managed to find film for it. Typical: right when he has a high-end camera on his smartphone, he finds a more expensive way to take pictures. While we're viewing the slides, my son is out in the garage looking for the old enlarger so he can set up a darkroom at our house. He resurfaces with a framed, dust-covered picture. An old trawler. *Arinbjörn Hersir* is written below the image. Grandpa Árni looks at the image and heaves a sigh.

"Ah, this picture. This is where it all started," he says. "Where it all started. Dad was on this ship."

Behind the picture we find a yellowed clipping from the newspaper *Vísir*, dated March 10, 1933:

> Shortly after *Arinbjörn Hersir* left port yesterday morning on a fishing voyage, one of the crew, Kjartan Vigfússon, fell overboard and drowned. Nobody noticed how it happened. After Kjartan went missing, the ship turned around and notified the police of his disappearance before going back fishing. Kjartan was married, 37 years old, the father of four children.[3]

Kjartan was Grandpa Árni's father. The story is told rather mundanely and the phrase "before going back fishing" shows that there was no reason to linger even if someone had fallen overboard. In the first quarter of 1933, thirty-four Icelandic sailors drowned and three foreign trawlers, with about forty crewmen, were lost at sea.

In 1933, Grandpa Árni was eleven. His father, Kjartan, worked as a boilerman on the trawler, and furnace work was both dangerous and difficult. Nobody knows why or how he fell overboard. I ask Grandpa Árni whether he remembers that day. He closes his eyes and finds an eighty-year-old memory there:

"I had cut across the ice-covered Reykjavík pond. I was halfway across when I realized the ice wasn't stable. The pond started to buckle under me. I started to sprint and the ice cracked every step of the way. I made it off but didn't stop running until I got back home to Óðinsgata."

That's when things fully cracked. Back home, a priest was waiting for him with the news that his father had drowned earlier that morning.

"I didn't know my dad," Grandpa Árni says. "He was always at sea and if he was ashore he was usually drunk. I don't really remember him."

It probably saved the family's life that those were the first years of a grand social ideal. They were given a newly built apartment in the workers' residence on Ásvallagata, thanks to the seamen's widows fund. These were modern, functionalist apartments designed by a twenty-three-year-old architect, Gunnlaugur Halldórsson, who had just brought over to Iceland the latest modernist trends from Europe. The apartments had running water, toilets, showers, and electric lighting; the residence was in a new neighborhood that the workers' union leader, Héðinn Valdimarsson, had had the initiative to build. The ambition was immense; a young doctor, Gudmundur Hannesson, was involved in planning the neighborhood with a focus on brightness and air quality. My grandpa benefited from the fact that in 1933 people had begun to understand how "the poor" could be as valuable as "the well-off." The workers' homes on Ásvallagata often came up in conversation with Grandpa Árni; he was truly grateful for having been given this apartment. He often asked himself what would have happened to them all if the family had been split up, or what their lives would have been like had they lived in a damp, debilitating basement.

Despite the apartment, Grandpa Árni still had to leave school at eleven to support the household, working as an errand boy and for a butcher: his mother was in poor health and his youngest sister had just been born. Some of his contemporaries at elementary school went on to become engineers and university professors.

"I was the best student in the class," he said. "Still, I think that those who went all the way to university aren't necessarily happier."

It's impossible to know. I think he lived his life as well and as carefully as possible, but I still detect some sadness and regret when he thinks about his schooling.

The workers' apartments also housed the Workers Library, full of books he would read in the evening after a long workday. He had a great competitive bent and was on the first team to win the Icelandic Championship in handball, Valur, right when people began playing this new game. He was also the Icelandic champion in the 1,500- and 5,000-meter races in 1942. In 2016 he was still sore about the sinusitis that had caused him to lose the 1943 race.

He saved up to buy a camera and soon became a passionate amateur photographer. He converted a basement closet into a darkroom and he could often be found running home from the track at Melavöllur Stadium so he could develop film to work out who'd won a close race—what's known today as a photo finish.

The slide machine hums and then there's a click and a bright light speckled with colors takes up the screen. Grandpa Árni adjusts the focus, bringing into shape rows of dahlias, endless dahlias he and Grandma Hulda had grown. Sometimes the pictures are upside down, sometimes they get stuck in the machine, sometimes two project at once. He adjusts the machine and then the flowers flow away in rows, pink and yellow and reddish dahlias, then mountains, cars, desert sands, endless snowdrifts up on the glacier Vatnajökull, pictures from Lech, in Austria, where they went skiing with friends each year.

The organization of the slides seems to have weakened somewhat, but he remembers everything in the pictures. He fishes up names and dates even though his short-term memory is spotty.

"It just happened one day," he said. "It was like a crack, like I'd been shot in the head with a rifle, all of a sudden I could not remember a single detail." He laughs when he says this; he is still very much himself. It makes you think how many traumas the brain can withstand before people cease to be themselves.

He has changed in recent years, though. He was so competitive when he was younger that we feared he might become a difficult old man, but he went the other way, becoming as soft as a wisp of eiderdown, grateful and almost sentimental when he praises Grandma.

When we celebrated his ninety-third birthday in November, he said: "What's going on? Is it Easter?" The next day he didn't remember the party. When I tell him he had a birthday, he laughs and says, "How fun! How old am I now?" I tell him he's 112. "No," he says disbelievingly, "can't be." Then I tell him the right number. He still thinks the number is far too high.

One time when he had to go to the hospital, I got into the elevator with him; looking at himself in the mirror, he asked, surprised: "Who's that?" "It's you, Grandpa Árni," I said. "No, I've got red hair!" "Not anymore, Grandpa Árni." "Well," he said, confused, "Some people called me carrot-top." Then he smiled. "But they only ever said it once," he added, and he showed me his balled-up fists and laughed.

He can't remember what happens from one day to the next, but he still remembers that someone talked me into running for president. "I think you're a bit too young," he says. "How would it be if you turned fifty and you'd *already* been president? What would you do then?" I wonder how memory works. He does not remember how old he is, he does not remember that he no longer has red hair, but somehow he manages to remember how old I am and can calculate how old I'd have been after two terms in office.

The slides continue to flick past on the roller blind. Now there's more logic to their order. There are highland trips, there's the house on Selás, the big white house Great-Grandpa Filippus built when my grandmother's family moved to the Árbær neighborhood during the war years, long before the suburbs sprang up all around. Next: a picture of my Grandma Hulda's siblings.

"I was so disappointed when my brother Thórhallur was born," Grandma Hulda says. "I closed my eyes tight and prayed, 'Dear God, take away his penis!'" She laughs and slaps her thighs. "But it turned out all right. Later, my sister Thóra was born."

"There were five siblings, right?"

"Five who survived. Mom lost two children," Grandma Hulda explains. "Two of my siblings only lived a short time. Valur, who was older than me, and Gudrún. They were born tiny and only lived for twenty-four hours. Great efforts were made to secure a priest because if they weren't baptized they wouldn't get into heaven. Then they just died; we never knew what was wrong with them. Then Thórhallur came along and he was just as frail as his siblings, but this time a doctor was called, and he lived. I often think about what would have happened to Gudrún and Valur if they'd had a doctor rather than a priest."

I think about the dynasty that has sprung from my grandma and her siblings. I imagine Valur as a great-grandpa somewhere in a big house in Árbær with sixty descendants. I imagine a historical novel based on his imaginary life, a four-generation story that begins when there is nothing but turf houses and scattered shacks. The story revolves around number 3 Selás, the big white house with the red roof. I suspect Great-Grandpa surveyed Bessastadir—the President's residence—when designing the house; it was almost as big and of similar shape. The family moved in in 1944, the year Iceland gained its independence. Back then, Árbær was thought to be way out in the country. Time passed and the whole neighborhood filled with houses, all of them perpendicular to their house. The book could be a kind of *Devil's Island*, a *One Hundred Years of Solitude* in Árbær.

"I never wore shoes," Grandma Hulda says. "Mom insisted I wear shoes on Sundays, and the rest of the time I'd walk along the banks of the Ellida river and talk to the *huldufólk*, the hidden people. One time, I managed to befriend a golden plover. I noticed her one spring, sitting on a nest. I was endlessly patient, going a little closer to her every day, until I could stroke her.

"We fetched water and did laundry in the Ellidaá. The laundry tub had a false bottom. Sometimes salmon slipped into it," she says, smiling. "We also got sent to the old turf cottage at Árbær to buy milk from the old farm lady, Kristjana. Where the built-up suburb of Breidholt is now, there was just a meadow."

She tells me she once saw one of the hidden people come strolling out of a cliff at the foot of the Breidholt slopes.

"He laid out his clothes to dry. I called my family; the whole lot of them ran out and saw him, but he simply evaporated before our eyes. We ran over to the cliff and the grass at its base was bent, but we couldn't see anyone."

Grandpa Árni picks up an album with pictures of Grandma Hulda when she was young; she is thirteen years old, pushing along a bicycle.

"This picture was taken in the summer of 1939, when I rode to Stykkishólmur to visit my grandma. It was the hottest summer I've ever experienced."

"Did you ride from Árbær?" asks my disbelieving son.

"No, actually, from Borgarnes, we took a boat there. We slept in barns and drank water from moorland pools along the way."

The journey would still have been some 150 kilometers, along mud paths. Grandma Hulda is the family hero; when we were young, we boasted about her, saying, "My grandma is stronger than your dad." She was the first Icelandic woman to get an airplane license to fly a glider when she took the test in 1945.

"That's a Schulgleiter," she says, "and after that I flew a Grunau Baby."

She shows me a flight certificate and a picture in which she's sitting on some kind of stick with wings. It most closely resembles the *Flyer* that the Wright brothers built during the early years of flight.

"Flying a glider is like nothing else. There's no engine, you're pulled up into the air and then you float along in silence, like a bird."

"Were you ever scared?" I ask her.

"You know, it's so odd, but I've never been scared," she

says. "Too bad parachuting wasn't around back then. I would like to have tried parachuting."

My grandma's brothers grew up dreaming of flight. They ran up and over the hill at Skólavörduholt to see the Graf Zeppelin aircraft soar over Reykjavík in 1930; in 1933 they watched in awe as the Italian pilot Italo Balbo and his squadron landed in the Reykjavík bay, twenty-four aircraft in total. Charles Lindbergh, the most famous aviator of his time, arrived later that summer; all these heroes were met with a royal reception and a brass band, and were invited to lunch with the prime minister.

Helgi, my grandma's brother, went to Germany just before the war to become a pilot. But he turned out to be color-blind and had to return home. It was a blow, but it probably saved Helgi's life or at least his conscience, because without a doubt he would have ended up in the German air force, the Luftwaffe. Great-Grandpa shepherded his children to the doctor and the dream of flight came crashing down when the doctor announced that all the brothers were color-blind.

When the volcano Hekla awoke from its hundred-year slumber and erupted in March 1947, Grandma Hulda phoned her friend who owned a small two-seater plane and asked if it wouldn't be fun to see the eruption. He landed on a frozen lake close to her house and Grandma Hulda ran to meet him. They took off and flew around the blazing volcano while ash tumbled onto the aircraft wings. The roaring volcano was so noisy they couldn't hear the plane's engine, much less talk.

"It was terrific," says Grandma, "the glowing volcanic rocks raining down. Some were the size of this table." She gesticulates wildly and laughs. "We landed in a meadow by

a farm, Ásólfsstadir, a short distance from Hekla, to have tea at the farm, but there was nobody home: everyone had fled!"

In March 1947, my mothers were barely six-month-old twins.

"What were you thinking, flying around a roaring volcano in a tiny plane with your six-month-old girls at home? How did your mom feel about it?"

"Dad always encouraged me," Grandma Hulda says. "He got a lot of pleasure out of my adventures. I knew the pilot, and what's more everyone at home loved to fly. If you're always worrying what might happen, you'll never do anything."

"Where was Grandpa Björn?"

She shakes her head.

"He was always somewhere out in the countryside attending to his medical duties. He rarely showed up and then he went to America to study surgery and I waited for him for six years. I didn't really do anything," says Grandma. "I hardly even went to the movies. Then he visited Iceland, bringing a car with him. I was so foolish that I went right ahead and took a driving test because I thought it was going to be our car. But he was only importing the car to sell it so he had money for tuition."

She tried on a wedding dress in anticipation of Björn's return to Iceland. Her father, Filippus, was more realistic. While she tried on the dress, he sat there in the living room calculating what Grandpa Björn owed them for child support. He was right. Björn had come to end the relationship.

"I was so thrown I walked loops around the pond in Reykjavík, devastated."

"Would you have wanted to be some doctor's housewife in America?"

"No, after thinking about it, I was lucky. I started hiking in the mountains and that's where I met Árni."

Grandpa Árni comes to his senses when he hears his name. He has begun to repeat himself lately, but his repeated words are mostly endless expressions of affection.

"Do you know how lucky I am to have got hold of your grandma? She's a real treasure. I don't know what would have happened to me without her."

"Where did you meet?" I ask.

"We met through Gudmundur from Middalur, in the hiking club. We were on the entertainment committee, and there was an annual gathering up at the ski lodge in Hveradalir. Then I invited her to join us on a glacial expedition. We needed someone to help us with the food while we were building the research base at Jökulheimar. Influential men from the Reykjavík Ski Society asked, shocked, 'Are you mad, planning to take a woman onto the glacier with you?' I had no doubt."

Grandma Hulda takes up the story: "I had only just joined the Icelandic Glaciological Society when I met Árni. I pestered him to take me out onto Vatnajökull; I had twice gone up Snæfellsjökull glacier, so it couldn't be more difficult than that."

The expedition set off in May 1955. The guestbook in the old lodge in Jökulheimar states:

This lodge is called Jökulheimar; it belongs to the Icelandic Glaciological Society.

We, the undersigned, built it as volunteers from May 30–
June 15, 1955, but we were gone, skiing on Vatnajökull
with Gudmundur Jónasson, the week of June 5–12. The
weather was generally favorable and we felt very content
here in Tungnaárbotnar and on Vatnajökull itself.

Jökulheimar, June 15, 1955

Hulda Filippusdóttir, Árni Kjartansson, Haukur Haflidason, Árni Edwin, Steinunn Audunsdóttir, Sigurbjörn
Benediktsson, Stefán Jónasson

Grandpa Árni shows me pictures of the lodge.
"We decided during this trip to carry out an even bigger expedition next year, as our honeymoon. My sister was
shocked: 'Are you going to take on a woman with two girls?'
she asked. I've never made a better decision in all my life."

Grandpa Árni is wearing a checked blue shirt, his hands still
strong and coarse, his corduroy pants worn from gardening.
His legs have failed him and I think of him as a precious
porcelain vase, an antique that totters so much I think he
will fall, but then he rights himself.

The pictures continue to appear on the screen and are
now mostly skiing photos. Grandpa Árni struggling with the
weight of the timbers at the lodge above Draumadalur up on
Bláfjöll. Because of the valley's name, the Valley of Dreams,
they called the shelter Himnaríki, Heaven. There are more
pictures from mountain lodges, of young folk singing and
dancing, pictures from the Kingdom of Heaven II, which
rose up after the first burned down.

"The name, Heaven, sometimes led to misunderstandings." Grandma Hulda smiles. "We were on the bus with my friend Magga, and her son asked, Where's Dad? He's in Heaven with Árni, she said. When will he come back? Monday, she said. A priest standing nearby nudged her and said with great sympathy but some zeal: 'You must tell the child the truth, my good lady!'"

In Grandpa Árni's collection there are documents of the first years of regular glacial research in Iceland, films of the first drills of the Air Ground Rescue Team, the first years of gliding, of skiing, of the Árbær suburb.

"The Heaven lodge was marvelous, we had a wonderful time. People often came to visit over the weekends and all Easter; we had seventy people in total up in the loft," Grandpa Árni says. "I reckon I can hardly count all the marriages that sprung from there during that time."

They were the first generation to "play" in the mountains. At that time, ski equipment was primitive; people had to walk several miles to get to the lodge in all weathers. There was no ski lift, no electricity, no Sno-Cats, no fleece sweaters and no GORE-TEX, no snowmobiles or jeeps to bring people to where they were headed, yet everyone who spent time at the lodge during that period talked about it as the pinnacle of happiness.

When you look back, it's as if that generation was faced with creating and constructing almost everything from the ground up. They had to found the Republic of Iceland, set up the social organizations, the theaters and orchestras, the rescue squads and sports clubs; build the houses, establish modern infrastructure, construct Reykjavík out of empty meadows. I'm sitting here with my grandparents, a strange qualm in my chest. As I listen to their stories and look at their pictures, I don't know if I'm trying to preserve their stories for the stories' sake or whether I'm trying to preserve my grandparents and their lives, to somehow mute an inevitable loss.

Grandpa Árni from Hladbær was born four years after the end of World War I. I think about everything he's experienced and about how the world can change in a lifetime. I think about all the wars and all the progress, all the revolutions in the arts and the sciences, and I think forward a hundred years into the future as I try to evaluate and understand scientists' predictions about where we are heading.

Grandpa Árni goes over to the slide projector when a picture gets stuck, leaving only bright white light on the screen. There are so many things I'd like to ask him before he passes into the light.

The all-encompassing silence
of God's great expanse

The old teak shelf in Grandpa Árni's writing office was buried under anthologies of flowers, books by Halldór Laxness, and the full collection of *Jökull*, the journal of the Icelandic Glaciological Society. It held one of my favorite books, Benedikt Gröndal's massive book of birds, and then one day I noticed a beautiful blue book on the shelf: *In Reindeer Country*, written by Helgi Valtýsson and published in 1945. The book describes Helgi's travels with the photographer Edvard Sigurgeirsson in the highlands north of Vatnajökull in 1939, 1943, and the year Iceland gained independence from Denmark, 1944. Their expedition sought the last reindeer herd in Iceland, the final descendants of the reindeer who had been brought to Iceland in 1797. Reindeer had once spread across the whole country, from the far north to the Reykjanes Peninsula, but had almost everywhere become extinct. The last herd survived in a kind of secret valley at the foot of a glacier, Brúarárjökull; the animals' calving grounds were in a place called Kringilsárrani.

Helgi was a romantic, a progressive, a poet. For some reason, his reindeer book had passed my notice at the time I was writing my own book *Dreamland: A Self-Help Manual for a Frightened Nation*, a defense of the Icelandic highlands and of the exact region his book describes. When I opened *In Reindeer Country*, I was struck by its lovely hand-painted photographs and by its travel descriptions. But the text was far from an everyday account of travels in the mountains. The words rose from time to time into a florid flood, a baroque hymn rather than a travel narrative:

> The wild highlands are a wide embrace, mountain blue. Their stillness quietens you, listening [. . .] Fascinated you attend to your own soul's breath, this essence you've forgotten about for years. It's here you first perceive your spirit's immeasurable expanse, and you stand still and astounded in the deep silence amid unspeakable reverence for your soul's divinity. The distance, the mountain blueing, the great glacier's dome, the weighty murmurations of silence — all this is reflected and echoed beneath your soul's vault, spanning Heaven and Earth, your spirit's wide horizon. You are moved to tears, resonant as a tremulous bell in the pregnant silence of God's vast expanse, becoming one with it.[4]

I read aloud, pausing for a long time after the words, *to be moved to tears, resonant as a tremulous bell in the pregnant silence of God's vast expanse.* I've read volumes of sublime nature poems but I can't remember having encountered something so elevated. This fragment was not an isolated example: the book as a whole is not just a travel narrative but a grandiose declaration of love for this nature. Helgi is writing a praise song for East Iceland and for the wilderness north of Vatnajökull, especially Kringilsárrani, which is unique in Iceland and, indeed, in the world. Kringilsárrani was once a kind of island, a fifty-square-kilometer wedge covered in vegetation, six hundred meters above sea level. It was bounded by a triangle formed of the Brúarárjökull glacier along one end and two untamed, almost impassable glacial rivers that merged on either side. There were high glacial peaks and so-called push moraines, heaps of fertile land the glacier deposited when it surged forward around 1890. It was almost as if the glacier had taken the surface vegetation and rolled it up in front of it like a carpet. These heaps could be as high as ten meters; they were unparalleled in modern geology. Only on Svalbard were there examples of glaciers that had thrust land up in front of them in this same way.

Helgi and Edvard stayed in this tremendous wilderness that hardly anyone had ever visited. Edvard photographed and filmed while Helgi kept a journal of their travels as he reflected on his confrontation with the mountain's glory. When Helgi left the area, there was regret in his breast:

Probably we fellows have said farewell to Kringilsárrani and Vesturöræfi for some time, and I perhaps for good. That thought peculiarly arouses a bitter sense of loss and mysterious longing in our breasts [. . .] Having once seen them, one does not forget these places' evening-blue, symphonic colors and contours, not if one has looked with the open eyes of body and soul. Unbroken peace, the stillness of wilderness, all beyond earthly understanding and of a higher intelligence—all these trickle and flow in over the rimy firmament of our minds and souls.[5]

Helgi Valtýsson was born in 1877 and came of age while the spirit of the first Icelandic independence movement resonated over those waters. Helgi describes his time in Kringilsárrani as a spiritual enlightenment. He writes his text into a hundred-year or more tradition in which true manhood involves composing praise poems about plovers and whimbrels and mountain lakes, a manhood that involves serenading the summer, the mountains, the flower-covered slopes, and hope itself. He pursued that despite living through other, harsher social realities of the times: infant mortality, poverty, disease. But this romantic view of life is notable for its gentleness; one is unlikely to find a better example of the unadulterated worship of nature anywhere else in Icelandic literature. Romantic philosophy arguably reached its apotheosis at that exact moment of the writing of Helgi Valtýsson's book. With the vocabulary of one hundred years of romanticism, Helgi could capture the impressions of the highlands in an overflow of baroque metaphors.

On their first trip, Helgi and Edvard were in Kringilsárrani for two weeks without any contact with the outside world, from the end of August 1939 into early September. By the time they returned, the Germans had invaded Poland, officially changing the worldview Helgi and Edvard espoused. The book was published in 1945, the same year the first nuclear bomb exploded, the same year modern "atomic poetry" emerged on the Icelandic literary scene. The world had lost its innocence; war's miseries had caused many to wonder where he'd gone, this so-called God. Books that unabashedly set their faith in a beauty that was in harmony with the almighty no longer paid their way. Poets instead composed inscrutable modernist poems. Steinn Steinarr wrote about "Time and the Water"; other poets wrote about the "nothing" that happened after death. Halldóra B. Björnsson wrote:

In earth's cold darkness our journey ends
and we no longer know this journey took place.

Europe burned, yet the war changed the world and gave rise to numerous industries. The aviation industry altered entirely, as did metal production; the nuclear industry emerged while mass production expanded, its capacity all too evident. The world's aluminum industry grew by 1,000 percent in just a few years in order to meet the war's need for bombs and airplanes.[6] The U.S. government instructed Alcoa to build twenty new factories in three years, giving them priority access to finance and raw materials.

At the end of the war, however, production did not slow back down again. The aluminum industry found an alternative outlet for its products with the emergence of the disposable consumer economy. Enterprising designers developed products that allowed people to use dishes, cutlery, food packaging, aluminum foil, and other valuable things only a single time. They packaged drinks in energy-intensive aluminum cans people could throw away instead of rinsing and returning as they did with glass bottles. This mindset went against the values held by previous generations, who had learned to respect what was valuable, to throw nothing away, to finish their meals, to fix things, to make use of everything.

The packaging industry and a consumer society thus combined to create an infinite demand for raw materials, which slowly but surely intruded into untouched areas around the globe. In 2002, the world's consumer machine stretched its tentacles to the northernmost corner of the planet: a decision was taken to flood the bulk of Kringilsárrani under a fifty-square-kilometer reservoir behind the proposed Kárahnjúkar Dam. The purpose? Producing subsidized electricity for the Alcoa aluminum smelter in Reydarfjördur. The plant produces a fraction of the aluminum that Americans throw into landfill. An annual amount of aluminum equal to four times the size of the U.S. commercial air fleet ends up on such trash heaps just from aluminum cans alone. Recycling cans in the U.S. would eliminate the need for three or four such factories.

"The all-encompassing silence of God's great expanse" and the whole environment Helgi described in his book finally drowned under two hundred meters of muddy glacial water when a plug was set into the Kárahnjúkar Dam in the fall of 2006. But the area did not drown forever because the water level fluctuates, revealing several square kilometers of shoreline of fine, light silt. Each spring, this lifeless land appears, gray as a ghost. Kringilsárrani is just one of the thousands of treasures today's generation of earthlings quite literally throw in the trash each year.

Writer's block

I went out to the Kringilsárrani reserve while it still existed and was able to experience firsthand this magical world Helgi wrote about. To walk through this valley of animals alongside the Jökulsá riverbed; to see goose nests amid the columnar basalt walls that formed the gorge's sides; to stand on the lava-red bedrock as the gushing glacial river surged over the falls of Raudaflúd—it was incomparable. How mesmerizing it was, watching the river break through a narrow crevasse and practically dissolve into a thundering torrent that flung rocks uphill, like some uncontrollable eruption, a formless disorder. A falcon hovered high above me. I found the lone rock standing on a fragile foundation, humanoid in shape, like a troll or Darth Vader's head. This rock had become symbolic of the area and it did not feel like any of the other rocks in the area. Even though I'd come in peace, I felt it was displeased by my presence.

I would never have dared speak my mind the way Helgi had, would never dare write something in his style: "There at Töfrafoss, my soul's harp strings vibrated while the Creator thrummed Kringilsá's burgeoning bass..." If I'd used this kind of language in *Dreamland*, I'd have been written off. I would have been held up as the archetype for some modern, New Age, urban hipster nonsense. I'd found myself overwhelmed by melancholy at the unruly devastation that washed out this peerless region, yet I chose words that seemed moderate and inviting to readers. I used the prevailing language of liberalism, innovation, utilitarianism, and marketing. I discussed the area's importance for Iceland's image, its potential tourist income, the area's research value, how the highlands were a magnet for foreign currency as a filming location for movies or commercials. Movies create an experience, but the image of a landscape can't replace a visit to the location. We live in times when money is the measure of reality. I couldn't argue nature's right to exist, its essential value, by saying that we might find God's all-encompassing expanse there.

This dispute over the Icelandic highlands flipped logical arguments on their heads. Instead of an area's untouched quality being taken as proof of its value and significance, that fact was turned against the region; as a result, it was branded "unpopular." Nature needed high enough "ratings," went the argument, people looking at it or using it, building hotels or gas stations or hamburger joints, bringing tour buses and guides. You had to be able to *use* nature, in some way, even if only as the backdrop for a car commercial. Nothing is allowed to have an undefined purpose; everything must be quantifiable, regardless of whether the metrics match the reality or not. The power to define reality and to discuss the value of nature belongs to economics.

Politicians talked down the region, saying it was nothing special. The farmers who "owned" the land went to the papers and sounded off about how the area really wasn't particularly remarkable, how it was primarily city folk who were losing their minds over the situation: "To be honest, I'd be happy not to have to go all the way up there into the ravines to round up the sheep each year."

I was part of the crowd arguing that the area had an excellent case for status as a national park, but I was often held up as an example of a modern urbanite experiencing nature for the first time. Similarly, photographers were accused of exaggerating its natural beauty via Photoshop to serve hidden agendas.

Helgi Valtýsson spent time in Kringilsárrani before any other interests entered its story. His book is the only independent environmental-impact assessment conducted in the region. He was simply someone who measured himself against nature's countenance; it certainly never occurred to him that humankind's destructive power could all too soon grab hold of this remote wasteland. "When I die, my soul will settle on Snæfell mountain," he said, and then comes the emotional eruption: "O God of Snæfell, everlasting spirit of the wilderness! You gave me this land as a promise, obliged me to love it and serve it with all the strength of my body and soul. Made it my yearning and my bliss, my sorrow and joy."

When I said the countryside was *beautiful*, I was accused of exaggeration. I was halfway through writing my fairy-tale book, *The Casket of Time*, when I went to Kringilsárrani. After walking around the area for a few days I found my ideas insignificant and pedestrian compared to the splendor that was imminently going to be flooded. I peered down at the Caterpillars, watching trucks and diggers erect a 190-meter-high dam, as darkly gray as the Death Star, across the Hafrahvammagljúfur canyon. I found it horrifying that fifty square kilometers of beauty should be sentenced to drown under a dead and gray man-made reservoir. The experience functioned as a kind of writer's block, filling up both my brain's hemispheres, allowing nothing else to come to mind. The Icelandic energy companies had big plans for the complete destruction of many of the foremost pearls of the Icelandic highlands. They were going to sink Thjórsárver, the largest nesting ground of pink-footed geese anywhere in the world, under a reservoir the size of Manhattan. The geothermal areas in the Torfajökull region were in danger: Aldeyjarfoss in Skjálfandafljót; the white, glacial rivers of Skagafjördur; almost everything that was beautiful and sacred in Iceland's highlands was at risk of being dammed, exploded, or drowned in order to sell cheap energy to multinational producers.

Iceland, which had escaped the worst of the industrial revolution, was hell-bent on making all the world's twentieth-century mistakes in the twenty-first century's opening years. Many people I knew couldn't write or think about anything else. You could see how the intense activism burned out one individual after another.

It was not until I read Helgi's book that I understood how my contemporaries and I were inextricably locked into the prevailing discourse. His writing wasn't circumscribed by an economic language in which education is an *investment* and nature is merely an untapped *resource*. The possibility that nature could be something higher, something more exalted, something beyond definition and even "holy" — in our time, that isn't considered a valid argument. Helgi was free, not required to discuss tourism, employment, export earnings. He was able to write the way he felt about beauty, nature, and the sublime.

When I was in this region myself, having spent a day in the crowberry meadow beside the roaring Töfrafoss waterfall before heading back to the campsite, a full moon looming over Mount Snæfell, I experienced what Helgi described:

> Quiet is the night and full of peace, fresh and pure, undisturbed. As though all physical connections to the outside world have been severed. The soul finds itself alone, brimful with calm and the night's all-encompassing stillness. Wordless and omniscient. The ages and eternity flow through her like a gentle murmur with an incogitable and sweet sensation of delight.
>
> In the tranquility of the wilderness, we understand fully what a glorious and amazing adventure our life is, a gift from God, one we seldom understand or pay proper heed.[7]

It says a lot that the area which was for Helgi the wellspring of sublime emotion and inspiration became for Icelanders the epitome of polarization, hostility, and disagreement. Those who experienced an awakening in the wilderness in the spirit of Helgi were labeled environmental extremists.

Helgi Valtýsson was born in East Iceland two years after the Askja eruption of 1875 caused its considerable damage. He came of age at a time when people still died of starvation in Iceland, when 20 percent of Icelanders left the country to seek opportunity in America and Canada. He was barely forty in the Great Frost winter of 1918, the year the Spanish flu killed nearly five hundred people in Iceland. In 1939, Helgi would have been in his seventies. Seventy years later, Iceland had an abundance of plenty. The most cars in the world per person, the most TVs, the most planes, the most trawlers, and by far the greatest aluminum production per capita. It was at that exact moment that the Kringilsárrani reserve had its protections removed and was drowned. The photographer Ragnar Axelsson called me when the wild rapids of Raudaflúd were submerged; he told me Studlagátt would start to go under by the week's end. I got a lump in my throat. I felt like he was sharing the death of a friend.

How could Helgi feel these sublime emotions toward nature during times when hardship and hunger still prevailed in Iceland? In Helgi's youth, romantic poetry was the most popular artform; in poems, generations of starving poets managed to praise flowers and birds with such art that half of Iceland's bird species have not been hunted or eaten since the mid-nineteenth century. According to Maslow's Hierarchy of Needs, those who have satisfied their basic needs ought to be better positioned for sublime perceptions of nature than the poor poets were in the romantic period. We've managed to produce more than triple the amount of energy the nation itself uses; we have our fill of food and our storerooms are packed with stuff. Why can't my generation speak freely, like Helgi, instead of being muzzled with imaginary economic terms and rationalist discourse? Shouldn't we be equipped to look past our feet and sense the larger context? It's as if those in power, prepared for every eventuality, have set things up by ensuring people never feel safe, that they always feel hungry and scared and are always ready to sacrifice one more valley, one more set of waterfalls.

That gives rise to a different question altogether. The sunken valley was about fifty square kilometers. Why isn't the reaction a thousand times more intense when we consider the global picture? According to scientists' predictions about global temperature increases, ocean levels will rise by between thirty centimeters and one meter this century due to the melting of glaciers and the swelling of the seas. All the Icelandic glacial melting combined will raise the sea level by more than a centimeter; if the massive ice sheets of Greenland and Antarctica start melting, we can expect a rise of tens of meters. If we are conservative and expect only a 0.74-meter rise over the century, about 400,000 square kilometers of land will sink into the ocean. That's four times the size of Iceland, and is an area larger than Germany.[8] There are big cities, coastlines, ports, and tidal flats at stake; these include the oldest cultural cities, world monuments, factories, summer resorts, farms, arable lands, nature reserves, and estuaries. About 115 million people live in these areas.

These consequences stem from rising sea levels alone, never mind the consequences of rising temperatures, desertification, drought, forest fires, falling groundwater levels, thawing permafrost, or ocean acidification.

I sense a buzzing inside me, the way all these words form a black hole I can't directly perceive because its quantity absorbs all its meaning.

Where Helgi found the all-encompassing silence of God's great expanse in one small area, what words might we use about the atmosphere we all breathe, about the way humanity is changing that atmosphere's composition? What words apply when people worry about the ocean's future, about its ecosystem? What words should we use for the rainforests given they are nothing less than the earth's lungs?

Should we draw words for discussing the Earth from science, emotions, statistics, or religion? How personal and sentimental can we get? Is it possible to use exaggerated expressions of affection, cold economics, military metaphors, complex philosophies? Right? Left? Beautiful? Ugly? Economic growth? Is the Earth an underutilized raw material, an infinite holiness, or must unspoiled areas be reduced to charts and graphs about the economic and social value of nature (see Environmental Assessment Report Appendix 4b)?

The same week in October 2018 that the United Nations Intergovernmental Panel on Climate Change issued a kind of "final warning" as part of a Special Report on Global Warming of 1.5 degrees Celsius, the Internet was consumed by an array of variously trivial things, all easy to understand and ripe for exciting hot takes. In Iceland, we were debating whether a Banksy reprint hanging in the mayor of Reykjavík's office was art or not.

The Special Report, which received little attention, dealt with all the world's oceans, its entire atmosphere, all the world's countries, humanity's hundred-year future, and the measures necessary to stave off disaster. It described in great detail the most rapid changes to have occurred in the climate at any point over millions of years. It explained their enormous impact on thousands of millions of lives over millions of square kilometers.[9] Sure, it mattered, but it was easier to have an opinion about a single art print in an office. In other words: all subsumed in a deafening buzz.

If my life is in danger, if my Earth and my descendants are in danger, aren't I obligated to understand what's at stake? What words manage to define the world?

Telling stories

My trip to Kringilsárrani became *Dreamland*, a book that carried me around the world. I gave a lecture in Munich and participated in a panel discussion with Wolfgang Lucht, a professor and scientist from the Potsdam Institute of Climate Impact Research in Germany. He said that not only had his most severe predictions from ten years ago come true, but they'd been exceeded. He'd never planned on becoming a doomsday prophet. His first love was poetry but he ended up a climate scientist because he was good at math. In his lecture, he talked about Greek mythology and about the curse of Cassandra: how it was decreed that she would be able to predict the future but no one would believe her prophecies. She was fated to know everything beforehand, and also doomed to see everything come true.

He said that after writing with such passion about landscapes, waterfalls, and hidden mountain valleys, I must be tempted to write about the most urgent issue of our times. I said that climate matters were complex and scientific; better to leave them to the experts.

"But you weren't afraid to criticize experts when it came to hydroelectric dams and aluminum smelting plants."

"True, but at least I could see the dam and walk around the land and understand where the reservoir was going to be. I was able to calculate energy production myself and work out how many unnecessary products the factory would make, and I could understand and criticize the engineers' calculations."

"Don't you trust yourself, then, to write about the biggest changes to the earth's most important systems since human-kind appeared — instead of wanting to make a handful of scientists responsible?"

"Can't they write about their own research?"

"No, because they're not experts in communication. Without help, their knowledge is destined to fall on deaf ears. If you're a writer and don't feel the need to write about these issues, you simply don't grasp the science or the serious-ness of the matter. Anyone who understands what's at stake would not prioritize anything else. I oversee a large team of scientists. We publish computer models and diagrams according to established scientific conventions; people look at them and nod and take them in to a limited degree, but they do not *understand* them, not really. I present data to parliamentary committees and explain how millions of peo-ple will lose their homes if we do not act. The politicians immediately respond: 'If we do what you say, hundreds of thousands of people will lose their jobs tomorrow.' They make it my responsibility. If politicians really understood what I was saying, people would roll up their sleeves and find solutions. We have focused such huge energy on deadly problems of war and weapons, or on reaching the moon. In the Manhattan Project, thousands of people were sent out into the desert to work through the nights, skipping summer holidays and Christmas holidays until they had created a nuclear bomb. So why can't we do something for the planet, for good? If politicians fully understood, they would come up with something like that. How many people should work on the climate crisis? Millions of people would not be too many when the Earth's future is at stake!"

I nodded. Perhaps I didn't look serious enough; I can't help but smile in the face of grave matters. So he said: "I'm not kidding, people don't understand numbers and graphs, but they do understand stories. You can tell stories. You must tell stories."

I thought about it.

"But no one wants to hear apocalyptic prophecies and dismal accounts of the world as it is today."

"That's the problem," he said. "Imagine a doctor not wanting to tell a patient he has early-stage cancer. That the patient needs to quit smoking immediately, to upturn his life, even put everything on hold for one or two years in order to save his life; to undergo surgery, radiation therapy, rehabilitation. Imagine if your doctor does not want to candidly tell you what might happen because he fears scaring you. So instead he recommends organic tobacco and peppermint tea."

"That makes sense."

"That's what's been happening. The result is we are faced with a serious problem that keeps growing: the patient has not changed their lifestyle, believing they will be saved by the scent of essential oils. We're talking about life and death, but people do not perceive it that way. Most of the solutions that get discussed are placebos. Homeopathic doses. Banning plastic straws. Sorting plastic. All minor details. We need much more radical action."

I listened to him, thinking how very serious he was about all of this. It's one thing to worry about a dam in the Icelandic highlands; does it make any sense to worry about the whole world? What kind of Pandora's box would it be to get involved in these matters, to cast one's happiness down into this infinite drain? There were upcoming conferences in Copenhagen and Paris and Rio and Kyoto. Thousands of experts publishing reports and charts. What more was there to add? Weren't politicians listening and responding?

Soon after that discussion, I attended a conference on climate issues at the University of Iceland where one specialist after another came up to the stage. A marine biologist talked about ocean acidification and seabird die-off. A glaciologist talked about glacial melt and an ecologist about the decline in global topsoil, about falling groundwater levels and the consequences of imminent water shortages. People threw out numbers, millions of people, millions of animal species, the fastest changes in millions of years. There was no agitation, no excitement. I looked around and the audience showed little reaction; the lectures might as well have been discussing the effect of agricultural tariffs on maize production. Shouldn't we have had tears in our eyes? Shouldn't we have divided ourselves into action groups and prepared a response that very evening? At the end of the proceedings, people packed up, chatted about this and that, and drove home like nothing had happened.

Perhaps we do not understand the world as individuals. Perhaps I was experiencing the antithesis of mass hysteria: a kind of mass apathy. Even an expert on the subject didn't seem to be able to breathe life into his research. He seemed unable to connect the deep experiences of diving and measuring the world's coral reefs to other people's imagination, to get across the sensations that arose from a knowledge of the impending death of everything he loved. Maybe scientists don't fully understand what they are saying until other people understand.

The words we do not understand

I understood that Icelandic has a
word for every thought on earth.
—Einar Benediktsson

We believe words are easily understood, that understanding them is natural to us, that the world we perceive and understand from newspapers and books is the world we perceive and understand. It's not that simple. Not at all. For example, we are accustomed to letting words like "global warming" pass us by while we respond to far less significant words. If we could perceive in granular detail what the words "global warming" contain, they should be like the threat in a fairy tale: we should feel terror. It can take decades, even centuries, to understand new words and concepts.

Pastor Hallgrímur Pétursson is considered one of the masters of Icelandic language and poetry. His *Passion Hymns* were first printed in 1666. The thirtieth hymn begins, "Hear, my soul, this sinfulness! Conscience should have intervened." The words in this verse—"soul," "sinfulness," "conscience"—these were the dominant words in that era's culture. For centuries, those words were pure power in the hands of priests and of the ruling class. Men confessed their sinfulness, purified their consciences, and secured their eternal souls in heaven. But these words had not always existed. During the ninth century, the Age of Settlement, it was unlikely the Nordic people would have understood Hallgrímur's lines. "Soul," "sinfulness," and "conscience" entered the language with the Christian faith around the year 1000. These words would not have registered for the enterprising Vikings. They robbed and pillaged without worrying about their conscience or about sinfulness. Men gained honor and esteem according to their raids; they were not to forgive their enemies but were instead obligated to take revenge. When they didn't take revenge, they might have felt a sensation that in some small way resembled a pang of conscience, but the word itself did not exist.

The poetry of the Vikings, skaldic poetry, was based on strict, specialized poetic conventions. In the wake of Christianity, a tenth-century poet would have faced a troubling challenge. Poems at the time were written using kennings derived from Norse mythology: they referred to Earth as "Odin's bride" and to heaven as a "dwarf's helmet." How could such a poet explain God as the creator of heaven and earth while writing in this tradition in which poetry itself was described as "mead of the Æsir" or "Odin's gift" or "Kvasir's blood"? Obviously, it would be problematic to refer to the pagan gods of the Æsir in a hymn praising the Christian God, the creator of heaven and earth. We use old ways of thinking to understand new ways of thinking: at first God could not be spoken about except in terms of the "dwarf's helmet" and "Odin's bride," relying on the very pagan worldview that Christianity intended to clear away.

Words affect our emotions, our feelings. Words enable us to get a handhold on the state of being and describe what slumbers in our chests. Words can tether actions that were previously invisible, frame them. In Icelandic, we have a word to describe a feeling of sweet yet melancholic nostalgia, the feeling that comes over you when you're listening to a meaningful, possibly sad song from the past. This word is *angurværd*, which you might directly translate as "tendersadness." The Faeroese have this concept, too, but their word is *sorgblídni*, which literally means "gentlegrief." These sister tongues, Icelandic and Faeroese, have adopted two pairs of synonyms to express the same sentiment: tender/gentle, sadness/grief.

I'm not sure if a mournful Faeroese feels exactly the same as a melancholic Icelander. But we could use such words to enrich our language and render the spectrum of emotions more precise. The tender remorse of *angurværd* might suffice to name the feeling that fills one's breast when people sing a tranquil song around a campfire. The gentle grief of *sorgblídni* is analogous, but in it there is a still deeper sorrow, a greater mourning. And so these two different words let us express our feelings with more nuance. I felt a strange *angurværd* when I looked at those old pictures with Grandpa; now that he is gone, my heart is filled with *sorgblídni*.

Hallgrímur Pétursson was born in 1614 and could write passionately about sin and grace, but he would have had great difficulty writing poems about freedom, human rights, democracy, and equality. He was a fine poet and a formidable thinker, but those words and concepts hardly existed in his century's language.

When, in 1809, Jørgen Jørgensen, the man Icelanders call Jörundur, King of the Dog Days, fomented a revolution in Iceland, he arrested the local Danish authorities and issued a radical proclamation. He said: Iceland is free, independent from Danish rule.

To our ears, this might sound like an obvious wish for a subjugated nation, a nation that had lost its independence in 1262. When I was in school, I learned that Icelanders had yearned for freedom for six hundred years. The reality was more complex. In all likelihood, nobody was asking for freedom back when Jørgen sailed ashore and issued his declaration. One fine day in the summer of 1809 a new and revolutionary idea set sail for the first time and came into being that very same day. But the problem was that no one had ever thought that Icelanders ought to be seeking freedom or independence. It's possible no one had ever spoken these words aloud in Iceland; as a result, they had little or no meaning.

Jørgen Jørgensen came to Iceland as an interpreter for a British soap merchant, Samuel Phelps, who was planning to buy tallow and lamb fat from Icelanders. While the countryside was well provisioned with those items, the war between England and Denmark had hindered sailing to Iceland for some time, and the country was beginning to sorely lack grain and other necessities. Count Frederich Trampe, the Danish governor-general, the Danish king's highest-ranking official in Iceland, tried to obstruct the transaction: Danish merchants had a monopoly on trade in Iceland, and contravening the law was punishable by death. Phelps and his crew detained Trampe and imprisoned him in a cabin on board the merchant's ship; meanwhile, Jørgen temporarily seized control of the country. He issued a declaration that Iceland was at peace with all nations. He made a national flag for the Icelanders and hoisted it: three saltfish.

At the time, strict rules governed domestic travel; Jørgen gave Icelanders freedom of movement, allowing people to move around the country as they liked and to trade without needing official documents or other permission; he mandated that every port engage in free trade with all states. That notion of free trade was a novelty in Iceland. He also announced that taxes would be cut by 50 percent immediately, putting an end to the practice of Icelandic tax revenue being exported to Denmark without Icelanders having anything to show for it.[10] In addition, he proposed that Iceland should always have a year's supply of grain as a way to protect against famine and economic fluctuation.

Jørgen had harsh words about the situation where a few "cowardly" merchants held the nation hostage. At the start of the nineteenth century, farm laborers constituted roughly 25 percent of the population, but as landless individuals they were without freedom, people who couldn't marry or have children, almost like drones in a beehive. Jørgen presented ideas for establishing a hospital and for ways to improve midwifery and prevent infant mortality. Over sixty summer days, from July through August, he managed to propose improvements in almost all areas of society.

He undertook to govern the country until the people elected a parliamentary assembly and established a republic. Article 12 of his Declaration of July 11, 1809, clearly states:

> That we declare and promise to lay down our offices the moment that the representatives shall be assembled. The time appointed for the convocation for the assembly is the 1st of July 1810; and we will then resign, when a proper and suitable constitution shall be fixed on, and

that the poor and common people shall have an equal
share in the government with the rich and powerful.[11]

The revolutionary spirit that had spread from France across
Europe had barely reached Iceland; the foundational writings that defined terms like "freedom," "equality," and
"independence" had not been translated or published in
Iceland. Jørgen was ahead of his time there, and almost in the
whole world. The Danes' constitutional democracy, established via the Danish Constitutional Act, did not come about
until 1849. When Jørgen stated in 1809 that the poor should
have as much of an equal share in governing the country
as the rich, his ideas went further than those of the French
Revolution did; there, franchise was based on property. At
that time, 88 percent of Icelandic farmers were tenants; the
idea that they were of the same stature as the well-off was an
absurd concept for most people. They felt themselves to be
lowlier, that power by default belonged in wealthier hands.

Jørgen wanted to give us *freedom*, he wanted to abolish the
monarchy in favor of democracy. He did not want power
for himself—he was an anti-monarchist—but the only word
the nation had to describe his role was "king," the same
way Norway's skaldic poets could not say that God was the
creator of heaven and earth without talking about "Odin's
bride." Icelanders scoffed at Jørgen, giving him the nickname
King of the Dog Days.

The apathy of the era's Icelanders disappointed Jørgen.
He offered people freedom, but no one understood what
he meant and so no one wanted to accept it. The idea that
a poor man could have as much to say as the wealthiest
man was completely at odds with their reality. People

had difficulty understanding how to run a country with a kingless representative assembly. A parliament amounted to a new concept, even though the medieval sagas spoke of Iceland as having a parliament, the Althing, with its system of administration by rural delegates—and it was not a given that people wanted to return to such a bygone system.

There were many good reasons to distrust Jørgen. He was only twenty-nine years old and an incorrigible swash-buckler, a gambler, a womanizer, and one might suspect his goal was to incorporate Iceland into the British Empire, but whatever his motives, the result was that radical ideas were being given voice in this country for the first time and people treated them like a joke. Magnús Stephensen, who was not only a leading judge but had also been instrumental in the founding of the Icelandic Society for National Enlightenment in 1794, offered the excuse in a letter that independence could not "be the wish of any good Icelander."

Those who would later struggle to achieve these notions of freedom, equality, and independence were either still in childhood in 1809 or hadn't even been born yet. Baldvin Einarsson, considered one of the founding fathers of the Ice-landic independence struggle, was seven years old; the poet Jónas Hallgrímsson was two; Jón Sigurdsson, Iceland's hero of the Independence movement, the man whose birthday is Iceland's national holiday, wouldn't be born until 1811—and his notions of freedom were still considered radical even by the time he reached middle age.

When the Icelandic parliament, the Althing, was reestab-lished in 1844, only property owners had the right to vote; that amounted to about 5 percent of the population. It was not until 1915 that disenfranchised men, as well as women forty years or older, gained the right to vote in Iceland. Men and women did not get equal voting rights until 1920. Complete independence—an end to home rule and the abolition of the Union with Denmark—was not achieved until 1944.

In elementary school, my generation learned that Icelanders had endured six hundred years of oppression under Danish rule, and that during that whole time the nation longed for freedom and independence. But that wasn't at all the case. It wasn't until the mid-nineteenth century that romantic poets came up with the idea that the nation had always wanted independence. Most people had been impervious to the zeitgeist, quite content to live day to day. People live inside their own realities, locked in the prevailing language and power systems of their contemporary moment. Most people think only in terms of the paths and concepts offered within a given era. It took Icelanders more than a hundred years of poems, speeches, forums, declarations, translations, and conversations in Copenhagen taverns to fully understand the terms set out in Jørgen's declaration. Only then was there a foundation for discussing these ideas and so a basis for achieving sovereignty in 1918; even then, there still had to be discussions about gender equality for the best part of a hundred years.

This book you are holding uses words that are as new to the language as the words Jørgen used back then. The term "ocean acidification" was only coined in 2003, by the atmospheric scientist Ken Caldeira.[12] According to the media registry Tímarit.is, this concept first appeared in print in Icelandic, *súrnun sjávar*, in the newspaper *Morgunbladid* on September 12, 2006.[13] After that, it appeared once in 2007, never in 2008, and twice in 2009. The word "profit," *hagnadur*, came up by contrast 1,170 times in 2006 and 540 times in 2009, according to the same source. By 2011, the debate had developed only so far as to warrant five print occurrences of "ocean acidification." "Kardashian" appeared 180 times.

Ocean acidification is an example of a concept that has passed us by, although the phenomenon is one of the most significant changes in our planet's chemistry and constitution over the last thirty to fifty million years.

What we are talking about is a fundamental change in ocean chemistry that could disrupt the entire ecosystem, a change so great that we might taste the difference in the ocean, with its pH level expected to drop from 8.2 to 7.9 or even to 7.7. The difference between numbers on the pH scale is logarithmic; most people thus struggle to realize how vast is the difference between each integer. It is a poor fit for the frame of reference inside our heads. We similarly struggle to grasp that a 4.0 earthquake on the Richter scale is one hundred times larger than a 2.0.

Ocean acidification stems from the seas having soaked up about 30 percent of the carbon dioxide mankind has released into the atmosphere. If we look at fluctuations in ocean acidity from twenty-five million years ago to the present, we will see a number of smaller fluctuations, some of which lasted over hundreds of thousands of years. If things continue as expected, the next hundred years will see a vertical plunge in sea acidity, as though a meteorite has crashed into the Earth. For the Earth, one hundred years is like a moment. For a process that once took millions of years to take place in a hundred years instead is a speed comparable to an explosion.

"Ocean acidification." I feel like I understand the words, but I probably don't. An empty gun looks like a loaded one;

a gun's usefulness, its harmfulness, depends on whether it's loaded or not. Words have different charges to them; it takes many years for concepts to reach full charge. "Ocean acidification" is as great and deep as all the oceans for all time. It is as vast as all the combined shoals of herring and sculpins, all the haddock and porpoises, the oysters, phytoplankton, and sperm whales; it is as massive as all the magnificent coral reefs with their turtles, brain corals, and clown fish. It is just as hard to swallow these words as it is a mouthful of sea butterflies.

If we examine the science behind ocean acidification and consider how many of Earth's inhabitants are reliant on the seas' health, we might wonder whether the full meaning of "ocean acidification" in 2019 is similarly weak as the word "holocaust" was in 1930 compared to its meaning in 1960. The term "ocean acidification" might become so significant that it will be future generations' dearest wish to be able to travel back in time and prevent the utter loss of paradise.

We inhabitants of Earth today are like the Icelanders during the time of the King of the Dog Days. It's as if the words "acidification," "melting," "warming," and "rising" don't elicit meaningful reactions the way "invasion," "fire," and "virus" do. We read the news and watch documentaries, but for some reason we keep to our daily routines.

Climate discussions are full of scientific concepts and complex statistics: 7.8 pH, 415 ppm. We must wrestle with aspects of chemistry, encountering words like aragonite, calcium saturation, and atmospheric carbon dioxide activity. We do not feel a connection with years like 2050, 2100, 2150, except when politicians make fuzzy plans to achieve a particular goal by, say, 2040. Politicians would like these issues addressed five to six terms of office beyond their own tenure. Countless "dog day kings" have come up with viable solutions that could benefit all Earth's inhabitants, but we have greeted them blankly, like a farmer in 1809 who has freedom placed in his hands but doesn't know what to do with it. And maybe they'll excuse themselves by saying that the forecasted end of the world was encrypted:

> 2100 is considered the year that aragonitic sub-saturation in the Arctic is expected to have a significant negative impact on calcium-forming organisms as ocean pH approaches 7.8 compared to the RCP 6.0 scenario outlined in the 2018 United Nations Climate Change Report of 2018.[14]

The message in this passage should incite fear, but for most people it is just jargon. Clauses like these should have a direct impact on politicians' policies, on voting in elections.

Jørgen believed that the whole public could be trusted to get involved in complex matters, to take a stand on them, and to vote for them. The world's peoples face a challenge. Scientists have pointed out that, based on current policies, we have paved the road to destruction. That puts our system to the test: Can we get so deeply involved in the issues that we elect people to power who can steer the world in the right direction?

Searching for the Holy Cow

In October 2008, I followed *Dreamland: A Self-Help Manual for a Frightened Nation* to the British Isles. The book had just been published in English and I was giving a public lecture in the small village of Frome in Somerset. I was staying at an inexpensive hotel in London; in the lobby, the newspapers had front-page stories about Iceland. On one was a cartoon of a volcanic eruption and a headline about the imminent collapse of the Icelandic banks. A young British MP who was supposed to be on a panel with me after my talk got in touch and withdrew. He did not want his name associated with Iceland. I went to the store and bought a beautiful sweater for my wife. While I was buying it, the Icelandic króna fell so sharply against the pound that the sweater's price rose by seven thousand króna in five minutes. When I returned to my hotel room, the phone rang; the woman on the line introduced herself as Halldóra and said that she had a somewhat unusual message. Would I be interested in interviewing the Dalai Lama?

That knocked me sideways.

"The Dalai Lama?"

"When he comes to Iceland in June next year. He's interested in discussing environmental issues."

I hadn't heard about his planned visit to Iceland and I replied, convinced it was a hoax:

"Great, really exciting. But I'm a Christian and I'll have to call the pope and get permission."

"Okay, do you want me to call back tomorrow?"

"Yes, I'm sure I'll have heard back from the pope by then."

The next day I got a call asking what he'd said.

"Who?"

"The pope!"

"Oh, yes, the pope. He simply said yes, amen."

"You need to send us your questions; they want to get them months before the interview."

I'm not a Buddhist and far from an authority on Buddhism. At that time, my knowledge of Tibet and the Dalai Lama was reliant on having read Heinrich Harrer's book *Seven Years in Tibet* when I was a teenager.

The Dalai Lama was born as a poor farmer's son in the Amdo region of Tibet in 1935. When he was two years old, colorfully dressed wise men showed up. They had followed the signs and the guiding stars and had brought treasures with them that had belonged to His Holiness the deceased thirteenth Dalai Lama. They presented the items to the child, who recognized all his old possessions. Subsequently, he was brought to Lhasa where he was began his monastic education at six years of age and took over the spiritual leadership of Tibet when he was only fifteen. In 1959, under the governance of Mao Zedong, the Chinese invaded Tibet and seized the country. About a million Tibetans were killed and more than six thousand monasteries were destroyed in the Cultural Revolution. The Dalai Lama fled with his immediate entourage to India and built a monastery in Dharamsala, at the foot of the Himalayas, where a small Tibetan village formed. Every year, refugees arrive; whole generations have now grown up outside Tibet, still hoping to return home one day. The year 2019 marked sixty years since the Dalai Lama fled over the mountains.

Of all the world's people, the Dalai Lama has probably lived the most extraordinary life and experienced the greatest change. He was born into the most closed society in the world, but as an adult became a kind of spiritual celebrity who discusses love, the environment, and Tibetan issues. For the Tibetan people, he is the Bodhisattva, the *enlightened being*. Almost like Christ was among us, not figuratively, but the same soul in a new body, an unadulterated continuation of past life and past personality. He is most holy and it is extraordinary to get an audience with him, let alone a whole hour, as I was being offered.

There can be few more distant places in the world than Tibet and Iceland. But according to Buddhism, everything is connected. What on earth can you say to a holy man who has been reincarnated fourteen times?

It was tempting to ask him about time, how our time is different from all the times the Dalai Lama has lived through, and how the future might possibly be more uncertain than in previous eras. The whole of nature. The whole of the future.

I searched for connections. I attempted to understand Buddhism, its division between various roles, with each more supreme and so on: Dalai Lama, Panchen Lama, Karmapa.

It is a complicated system. Buddhism has no god, but it is full of folklore, demons, supernatural beings, invocations, prophecies, superstitions, traditions, and holy ordinances. How to address the holy man? Heinrich Harrer met the Dalai Lama when he was eleven, made friends with him, and later wrote *Seven Years in Tibet*:

> We were told that the name Dalai Lama is not used in Tibet at all. It is a Mongolian phrase, meaning "broad ocean." Normally the Dalai Lama is referred to as the "Gyalpo Rinpoche," which means "Treasured King." His parents and brothers use another title in speaking of him. They call him "Kundün," which simply means "Presence."[15]

Reincarnation is a curious concept and in conversations I've had it seems to me to be quite literal. People talk about a reincarnated person and his predecessor as the same person in different bodies: I knew him in his previous existence; back then he was serious but now he is energetic and really funny.

My son, in fact, had aroused my interest in ideas of reincarnation. When he was three years old, he told me how he was born. I'd been sitting by the fire roasting olives, he said, and I'd gone with him to meet mom. You were so happy when you saw her, you kissed her, he told me. He also told me about his "old mama" who had "died some long years ago." He also told me about his "old brother," who died when he got hit in the forehead with a rock. Later, on a bike ride, as he was riding in a child's seat behind me, I asked him again about his "old mama" and he said, "I can't talk about my old mama anymore." After that, he stopped mentioning it. What do I know?

I read books about Chinese history and Tibetan history and the Buddhist holy texts. I read reports, articles, pored over maps, digested biographies and films. I read books by or about the Dalai Lama. Some good, some quite unexpected: *The Art of Happiness at Work.* Pearls of wisdom for middle managers. I read books about happiness, marriage, youth, and the future. I read about people imprisoned for having a photo of the Dalai Lama in their possession and about the six million Tibetans living in subjugation. I felt a responsibility toward them. I could see how human rights had been systematically violated in Tibet and how the chief institutions of Tibetan Buddhism were being sundered. The Panchen Lama is the second-most important office after the Dalai Lama; his reincarnation was recognized in a six-year-old child in China in May 1995. Three days later, the child disappeared with his family and nothing has been heard of them since.

I read *Dhammapada: The Way of Truth,* and the simplicity of the text was beautiful; here and there it was like Hávamál, the Icelandic poem "Sayings of the High One":

> Watch out for duplicitous talk
> and limit your own words.
> Avoid rude speech
> and practice kind conversation.[16]

I puzzled over mythology and, since the Dalai Lama was in India, I thought about sacred cows. Why do we Icelanders find holy cows so alien when our own mythology depicts the source of life itself as a sacred cow? According to the Prose Edda, the world begins with the cow Audhumla, who was created from hoarfrost:

> Hár replied: "And right after the rime dripped down, a cow called Audhumla was suddenly there, and four milk rivers ran from her udders, and she gave birth to Ýmir."
> Then Gangleri said, "What nourished the cow?"
> Hár replied: "She licked salty blocks of ice. And as she licked the ice blocks, that evening the hair of a man emerged; by the second day it was a man's head. He was called Búri. He was handsome, mighty, and powerful."

Audhumla nursed Ýmir, from whom the world was created. His blood became the oceans and waters, his flesh the earth, his hair became forests and his brain became clouds. Marvel Comics did a whole series about Thór, Ódin, and Loki; operas have been made about Valhalla and Ragnarök—but what about Audhumla? Even her name is a mystery: *aud* means "prosperity" in Icelandic, but *humla*? No one is sure. Her story sounds like some ancient fragment of a fragment of mythology, a remnant, like a game of whispers that has gone awry over several thousand years of traveling across two continents until it became: In the beginning there was nothing but Ginnungagap, the bottomless abyss, and then came a frozen cow who fed the world with four milk rivers . . .

As I kept reading, I found out that Audhumla has a sister in India. The mother of all India's cows, according to Hindu lore, is Kamadhenu, a cow of abundance and plenty, closely related to the sacred Pritvi, which is the Earth itself, often depicted as a cow. Kamadhenu is derived from a similarly vague abyss as Audhumla. All the gods have a refuge in her body; her eyes are sun and moon deities but her feet, her very foundations, are represented in images as mountains, as the Himalayas.

In the Icelandic poem "Helgi Hundingsbana," found in the Poetic Edda, you find this passage:

> In ancient days
> eagles cry out,
> holy waters fall
> from Heaven Mountains.

In Icelandic, "Heaven Mountains" is *Himinfjöll*, which sounds like *Himalja*, the Himalayas. I searched for additional connections and saw that in Nepal there is a region called Humla. The great Himalayan trail runs from there: the old salt road that goes up to the sacred Mount Kailas in Tibet. This mountain is axis mundi, the center of the world, the most sacred place on earth according to the ancient philosophy of Buddhists, Hindus, Jainists, and Bonpa. It is sometimes called the stairway to heaven. The sun and moon revolve around Mount Kailas, home to the throne of the god Shiva, who is said to sit atop a great bull.

At the foot of Mount Kailas is Manasarovar Lake, long considered the highest lake in the world. From this region originate four of the most sacred rivers in Asia, flowing in all cardinal directions:

Indus is one of the longest rivers in Asia, about 3,200 kilometers long; it flows from Tibet, through western India and into Pakistan.

Sutlej runs about 1,500 kilometers from Tibet through India and Pakistan, where it merges with Indus.

Brahmaputra travels through India and Bangladesh, a total of about 3,000 kilometers, before joining the Ganges at the Meghna estuary in the Bay of Bengal.

Karnali flows through the Humla region, gets the name Ghaghara as it flows through Nepal, and forms one of the largest tributaries before finally becoming tributary to the Ganges, 1,000 kilometers long in total.

One of India's holiest sites lies at the foot of the Himalayas, in Uttarakhand state. There, one finds a valley glacier by the name of Gomukh; from under this glacier, foaming white, breaks one of the most important headstreams of the Ganges. "Go-mukh" literally means the mouth or face of the cow. "Go" is cow. "Mukh" is mouth.

Suddenly, everything seemed to click, all at once, like a blurry origin story from Nordic mythology was pointing out the most sacred mountain in the world, the glaciers of the Heaven Mountains, the source of Asia's chief and most magnificent rivers.

A cow made from rime is the perfect meatphor for a glacier. A glacial river contains no ordinary water at all: it flows milky white with dissolved minerals because over millennia the glacier has ground the rock face beneath itself.

Glacial water is one of the best fertilizers that can be found for fields and meadows; these sacred rivers are the life source for Pakistan, Nepal, India, Bangladesh, and China. In total, more than a billion people rely on the sacred waters that descend from the Himalayas. Much of the water originates in thousands of glaciers, some of which reach more than 7,000 meters high.

In Sanskrit there are words that sound like Humla. *Haimala* is winter and *hima* means snow, frost, or rime. Audhumla. Ice as the source of life and prosperity. Buddhism assumes that everything is interconnected and suddenly I find all the threads coming together. Could it be that there is some connection there? Actually, there is a word in Icelandic (*samband*) that's also in Hindi (*sambandh*): they both mean "connection." Everything is related to everything. Everything fits.

You'd have to spend your life working in comparative studies to coin a proper theory about these links. But that's not really the point; I'm mainly looking for poetic echoes, and Audhumla—the frozen cow—provides a perfect encapsulation of the role of the Himalayan glaciers, and glaciers all around the world, for that matter. The Indus-Ganges plain is considered one of the cradles of civilization. People settled by these milky rivers, began agriculture, and domesticated animals. The glaciers are a perfect system: they collect water when monsoon rains pour down, then supply meltwater to densely populated areas in the dry season; in this way, they deliver vital and nourishing glacial water. The glaciers compensate for extreme fluctuations between monsoon and drought periods and during the hottest times glacial water is

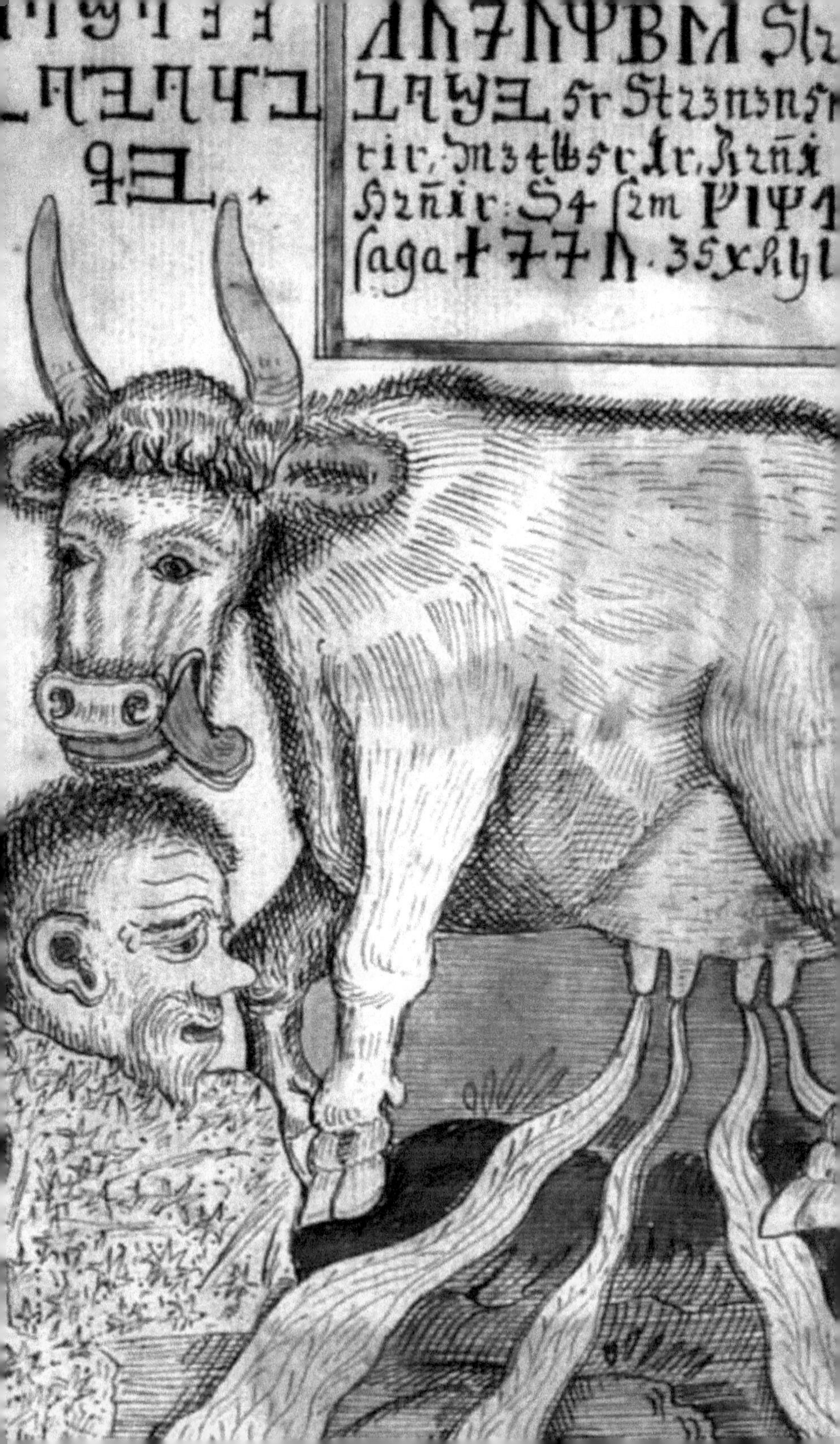

sometimes the only water available. This system maintains groundwater levels in large areas and is crucial for crops and vegetation. In the places where monsoon rains don't fall, glacial water is of utmost importance; it can constitute up to 90 percent of the water available to people.

How did Audhumla come into existence and how did she manifest as a mythical memory in a manuscript on an island up north just below the Arctic Circle, two hundred years after Icelanders became Christian? There's no way to trace the story across its eight-thousand-kilometer journey, but we are all connected, we all come originally from the same people. The Prose Edda says that the gods of the Æsir came to the Nordic countries from Turkey, whereas the explorer Thor Heyerdahl thought they originated in Azerbaijan. The relationship between Indo-European languages is still evident many thousands of years after two sisters led two cows along, each heading in her own direction taking her people and their stories. It's possible one of the sisters was named Edda and the other Veda.

From a cow's-eye view of our culture, it's likely that those who settled in Iceland from Norway and the British Isles could trace their roots to Indo-European cattle farmers. The unit used to measure prices and value during the Icelandic Commonwealth, 930–1262, the age of the Sagas, was the price of cattle; human worth was assessed in the value of cows. In the medieval Icelandic *Book of Settlements*, in the Hauksbók manuscript, there's a description of how a woman should settle land:

But it was so commanded that a woman should not settle a broader expanse of land than she is easily able to
lead a two-year-old heifer or an adolescent bull around
between two sunsets on a spring day.[17]

The letter *A* in our alphabet is thought to have its roots in
aleph, a symbol that was an ox's head in Egyptian hiero-
glyphs but, turned upside down, became the Greek *alpha*,
related to *alif*, which in Arabic means tamed. The letter *B* is
considered to have been a symbol for a house. The tamed
cow, the permanent dwelling; the symbolic *A* and *B* of
human history. In many places, cows were what allowed
us to set down roots, construct houses, plow fields, sustain
households, build villages, and raise cities, systems, and
religions that in turn often acknowledged cows as a symbol
of the source of life, wealth, and happiness. Cows could
transform grassland into food and settlers leading a cow
behind them were able to settle anywhere so long as they
could make hay for winter fodder. Anyone who can milk a
cow will never starve. And thus a person could settle down
and raise a small family wherever grass sprouted and water
flowed—and from there people looked up at the mountains,
at the great spring the water came from. White Audhumla
with peaks like horns, the source of life, the Himalayan Universe Cow. Some believe that cows became sacred in this
part of the world when overpopulation arose: these most
densely populated countries could never support so many
people if they ate meat all the time.

Everywhere you set foot in the Himalayas, you can find
connections and correspondences between glaciers and

cows. There are ice caves where the sacred icicles are called "udders." We say that glaciers "calve" when ice floes come loose and break off the ice sheet into the sea or a lagoon—the same word used for cows when they have offspring. Why do people say glaciers have calves?

I was elated when this correlation dawned on me. I felt like I had made a historical discovery and I rang my friend and mentor, the natural scientist Gudmundur Páll Ólafsson, to tell him about it. There was silence on the other end of the line. For a moment I thought I had gone overboard with these fanciful parallels. Finally, he spoke up: "I'm going to send you a chapter I finished today for my forthcoming book, *Water in Icelandic Nature*." He sent me part of the magnum opus he was writing. It started: "The Himalayas are the world cow and Mount Kailas, which milked Audhumla . . ."

He too, then, had had a revelation about Audhumla. I found it almost mysterious. How had we each separately conceived this? At that moment, there were about two billion people on earth who believed in the sanctity of Mount Kailas; someone long ago should have seen this lyrical parallel. Why was Audhumla mooing now? We decided to share the idea.

Gudmundur Páll never saw his book about water come out. He died of cancer in late summer 2012, aged 71, and the book was published posthumously. The *Nepali Times* published an obituary written by Kunda Dixit. "He became like a guru to me," the Nepali author said. I felt the same way. Gudmundur Páll dreamed of a society where people could balance technology, culture, and nature. Where people could use modern science to deepen their understanding of nature and live in harmony with it instead of plundering it.

Before he died, we talked about faith and life after death. We were convinced that if he came back he would come as a bird, not as a majestic eagle or a great northern diver, but as a misunderstood bird like a starling. Gudmundur Páll always took the side of misunderstood animals. Two days before his funeral, I opened the door to my office and heard a strange whistling of wings and a chirping. A little starling had somehow come in and was beating its wings frantically against the window. I could hardly believe my own eyes. I carefully caught the bird, held it gently in my grasp, and my eyes filled with tears. I opened the window, released the bird, and watched him fly out into the late-summer sun.

Today, you can observe rapid changes in snow cover in the highest peaks of the Himalayas. This happens several thousand meters above sea level where people thought they would be unaffected by climate change. The creeping valley glaciers have retreated and left behind them unstable glacial lagoons that sometimes burst and sweep away towns and farms downstream. In some places, the glaciers have thinned by one meter a year, and scientists point out that such thawing will directly impact the lives of tens of millions of people.[18] While the glaciers are melting, river flow temporarily increases. The volume of water increases as the ice melts. It's one thing to react to rising sea levels, but what to do if a water source, the flow of life itself, ceases? Is Aud-humla dying?

I had enough big questions to bring to the table for His Holiness the fourteenth Dalai Lama. Not to mention all the other topics he's contemplated: love, friendship, hope, peace, the future.

A visit from a holy man

June 2, 2009

One of the world's greatest spiritual leaders is coming to Iceland and at the same moment the country's leaders are fleeing. The president has gone to Cyprus, to the Games of the Small States of Europe; the foreign minister has an unexpected meeting in Malta; and the prime minister does not have time to meet with His Holiness. Despite his peaceful message, the Dalai Lama has caused considerable upheaval in his travels around the world and Chinese authorities have taken a firm stand on states that formally welcome him. Björk whispered "Tibet Tibet" at a concert in Shanghai and subsequently the rules on foreign musicians' visits were tightened.

The demands of the Dalai Lama are modest: that the Chinese go "midway," allowing Tibetans to cultivate their language and culture, with voting rights and a voice inside the Chinese state.

I thought the flight of our authority figures worth reflecting on: What is the purpose of Icelandic independence or democratic nations in general if such nations do not stand by the rights of the weak in the face of the strong? Should we be a cowering mouse or do we have a duty to support the oppressed? On the other hand, China is a complex phenomenon, home to one-sixth of humanity. There is no way to serve the country, the nation, or the government with simple statements. Generalizing about China and the Chinese is almost like generalizing about humanity.

I meet the film crew recording the interview on the seventh floor of Hotel Nordica in Reykjavík. I run through my questions in my mind; I had sent some over in advance and I was wondering how much I could bend them. There were many things I wanted to ask aside from questions about the environment and the melting of the glaciers.

His Holiness comes in, smiling. He blesses me and gifts me a white cloth. Arnar, the cameraman, attaches a microphone to him; His Holiness tugs Arnar's huge red beard and laughs. He's wearing a maroon robe, his hands and shoulders bare; when he laughs, his whole face joins in. He speaks a distinct, simple English and when he falters, he asks for help from the interpreter or his assistant.

I start hesitatingly, greeting him warmly, thanking him for letting me interview him. I first ask whether he didn't feel overwhelmed, having the whole world on his shoulders when he was just a child. For most people, being a human being is enough of a task.

He squints his eyes slightly and responds:

"I became the Dalai Lama unexpectedly. The responsibility during those years was very limited. I only had Tibet on my shoulders."

"Only Tibet," I repeat. The country was roughly 2.5 million square kilometers in size.

"Yes, and since then, I think a hundred thousand Tibetans have become refugees, including myself. During the period

I've learned many things. I've also noticed problems on this planet. Many problems are essentially man-made. Many materially well-off people, even billionaires, are very unhappy.

"Then there are the ecological problems. I am concerned about those problems. Wherever I go, I speak on a human level. Not only on the level of Tibetans. I speak about inner values, like warm-heartedness. We are social animals and given today's reality, we need a sense of global responsibility. The old notion of 'we' and 'they,' that clear demarcation, without concerns for anyone but oneself, is outdated, it's unrealistic. Because this world is heavily interdependent. We are all interdependent. That's the reality."

I'm keen to ask him about Audhumla. I guide the interview in an environmental direction and ask him about the melting of the ice.

"I've been living at the foot of the Himalayas for the last fifty years and we have experienced great changes over the last forty to fifty years. In the early period, there was a lot of snow. Now, decade by decade, it gets less and less. Even where I live in Northern India, what will become of the water supply in a few decades? What will happen? Already we are worrying. The same thing is happening in Tibet."

I say, "In mythology, ice is often associated with death. But now when we look at the Himalayas, we suddenly understand that ice is the source of life. And the water flows under the glacier white as milk."

"Hindus don't even call the water of the Ganges water, but nectar. That's how they describe water so sacred and pure it's simply called nectar. Since I was there, I express it a little differently: whether or not you consider it nectar, in reality it's water! Ha ha!

"According to many people I've talked with, many century-old glaciers are shrinking and temperatures are increasing. One Chinese ecologist states that global warming is on average 0.1 degrees; in the Tibetan plateau it is 0.3. So much faster.

"Specialists describe there being a North Pole, a South Pole, and a Third Pole. The third pole means the Tibetan Plateau. Some of these scientists are of the view that, in its effect on global warming, the Tibetan Plateau is almost similar to the North and South Pole. The Tibetan Plateau has a very high altitude and a cold climate, so nature's ability to restore itself takes longer than in a warmer climate. Once you damage it, it takes a longer period to recover. We need to take special care."

"In recent years, there has been much talk about rising sea levels, but much less about related issues, how billions of people have built their livelihoods on the glaciers and on the glacial water flowing from Tibet and the Himalayas. That brings us to something I wanted to talk to you about. In Nordic mythology, the world begins with Audhumla, a cow created from hoarfrost. From her teats come four milk rivers which feed the world. The story seems to have a remarkable similarity in Tibet and Mount Kailas."

His Holiness looks at me. He whispers to his interpreter and then looks back at me and starts laughing.

"Magic cow!" he says. "Magical cow! Mount Kailas: I've never been there. Of course, spiritually, people consider the mountain sacred. And I think four rivers come from Mount Kailas. Mount Kailas is very sacred and I sometimes jokingly tell my Indian friends: the Indians' god, Lord Shiva, his permanent residence is Mount Kailas. That's inside Tibet.

From that viewpoint, millions of Hindus' god is actually a Tibetan god. Then you have Tibetan Buddhists: our master or teacher, Buddha, was Indian. I sometimes joke like that. Kailas, though, is a very important mountain. The major rivers that cover the whole of Asia, from China to Pakistan, come from the Tibetan Plateau. Tibetan nature concerns not only the six million people living in Tibet but the billions of people who rely on the water that comes from there. It's not just Mount Kailas; it's the entire Himalayan snow mountain range. From the Chinese border up to Afghanistan. This extensive snow mountain range. Many scientists and ecologists predict that in the next twenty to thirty years, some major rivers will be reduced, and some rivers may dry up. Before that, because of more melting of snow, there will be more floods. But then eventually it will dry up. So very serious. Usually I feel, and also tell people, that unless we take special care, if the present tendency continues, then I think in the next generation a billion human beings will suffer that threat.

"It's not only the melting of glaciers that's a problem. Forests are also suffering. In some ways, the Chinese government has taken steps to protect the environment, including stopping deforestation. But as everybody knows, there's a lot of corruption. Even if companies put some sort of regulation or policy in place . . . there's bribery, these things keep happening. So it's a very serious matter."

"We live in a time when, incredibly, humans even have the power to damage glaciers. Are we living through a mythical time, in your view? Do you think today's issues are more serious than humanity has ever seen?"

"Yes, that's true. I think the whole universe, over millions of years, is changing. The world's position is changing. And also the heat of the sun. An unbelievable amount of hydrogen is burning, according to scientists. I think our sun is comparatively young, five billion years old; its future is another five billion. In that way, even the sun itself is changing. Then the whole universe around the sun is changing. That's its nature. We can't do anything about it.

"But according to reliable experts in ecology, this rapid changing in global warning—there is definitely a human contribution. Too much deforestation in the Amazon and also in the Himalayan range. In Southeast Tibet, which has a border with Burma, there is one of the thickest forests on earth, and a lot of deforestation took place there. Then, of course, cars and factories. Coal combustion has caused a lot of damage.

"If people worldwide would pay more attention, or show some sort of effort, I think the warming can be postponed a little. It's a question of not only a few people and a few countries. It's a question of the whole Earth. Whether seven billion humans on Earth will survive or not depends on it. If the whole world becomes a desert, everyone will have to go. I don't think anyone wants that. Unless we take the fullest precaution, our children, grandchildren, will face immense consequences. And it will by then be beyond their capacity to do anything. Therefore, I think our generations have great responsibilities. That's how I feel."

"But do you think we can achieve these goals peacefully?"

"I think so," he replies. "I'm not the expert. My own little contribution is that I usually save power and save water. Wherever I go, I never take a bath, only a shower. Whenever I leave my room, I always turn off the light. I consider that my little contribution. I think it's important that everybody does that. Of course, many people are without water and electricity; that's a different question. But those people in more developed areas, I believe a responsibility for ecology should be a part of everyday life. When it comes to big industry, I think they have a bigger responsibility. And their work has more effect.

"This is a serious matter for China, a big populated nation which already has water scarcity. There has been some talk of diverting a river that runs from Tibet to India . . . they talk about turning it to China. This has some benefit for that area, of course, but immense consequences for their southern neighbor.

"And also pollution, due to a lack of awareness. As I said, the Chinese government is now realizing the importance of these things. Since this is a global problem and if something is wrong in the Himalayan region, I think it will affect Iceland and North America. I have heard say that you Icelanders have a lot of glaciologists. So some of your experts should go to Tibet with the help of Chinese scientists and carry out some thorough scientific research on how much is already damaged and what is the best way to take certain measures to prevent these things."

"But what about Tibet and the Tibetan nation? It isn't faring well."

"In the long run, China will have a more sensible regime, one that is a more open, more transparent, without censorship. Then the reality of the Tibetan situation will become clear, particularly to the Chinese people. We can already see that. Many Chinese intellectuals and writers are already expressing their concern about the Tibetan situation and are critical about their own government's policy. These people are not pro-Tibetan or anti-Chinese. These are normal, educated people aware that the present situation is neither good for Tibetans nor the People's Republic of China as a whole. But the Chinese masters are completely ignorant. Because of censorship. China is the most populous country in the world, with 1.3 billion inhabitants, an ancient nation now doing well economically. China has potential to play an important role on this planet. For that reason, trust with the outside world is essential. Too much censorship, without accountability, means trust is difficult to develop. In the long run, once China becomes more democratic and a more open society, with free speech and, particularly, freedom of the press, then, I think, things will be much healthier. Much more peaceful."

"I live in a European country with a free press where we should be able to speak our minds in public. Yet we can see that European leaders are frightened; for example, they are afraid to meet you out in the open. Your fight so far has been peaceful."

"The very nature of my visits, including here, is non-political. The purpose is mainly spiritual and educational. My number one commitment is to promote human value in order to have a happier human society, a harmonious society. In another sense, as a religious person, to try to promote a close understanding across different religious traditions so that genuine harmony can develop. That way, various religious traditions can make some kind of effective contribution to common issues, environmental issues and human values, human rights. That's my main concern. If in the meantime I have some opportunity to meet with some leaders, I'm happy. If they find it a little inconvenient, I don't want to create inconvenience. But these things, promoting human value and religious harmony, mainly depend on the public. On the people. Wherever I go, I always have some public meeting. To consult with people and with the local organizers. For that, I feel very happy and that's what's most important."

"Do you see a link between people's spiritual feelings and how we treat the planet and other people?"

"I think we are the same, people of snowy lands. You Icelanders are also from a snowy land, so naturally there is, I think, at some mental or emotional level some similarity. Iceland is a small country, a small population. I think that's fortunate. The Netherlands is also a small country, but densely populated. I think I prefer your island, although I think in summer the day is too long and in winter the night is too long. A little out of balance! Otherwise, I think your country is beautiful. Very beautiful."

"A good place to be reborn, perhaps?"

"Yes. Ha ha! Why not?" he says, laughing.

"You'd be very welcome."

"Thank you! I think it's important that Icelanders are a part of Earth's seven billion. We very much appreciate your voice about Tibet, your concern. The Tibetan issue is not just political; I consider it also a moral issue. The present condition in Tibet, this crisis, is neither good for Tibetans nor China. In the interests of Tibetans as well as China, a proper solution must be found, a mutual solution."

"We Icelanders are very distressed by our economic crisis. You might not notice the crisis as you drive through the streets. Maybe it isn't a crisis. Do you think crises can be beneficial?"

"I think that depends on one's own attitude. Those people

who always consider money to be important, always thinking about money, even in their dreams, I think those people get the most disturbed by financial crisis. In their minds, financial problems are a disaster. Such problems affect their whole lives, their health. Of course, everybody considers money important. But there are other values, such as a happy family, compassion, some prayer, and some energy or time spent in some other field. I think that people who are mindful of these values experience global economic crises as less disruptive. That's how I feel. This crisis might remind us to keep other values visible, not only money. When material value is the primary concern, there are limitations. People must look at some other thing that provides them with peace of mind, satisfaction, a joyful life."

"What about the meaning of life? Have you found it? For yourself?"

"According to my own beliefs or experience, one's own life, if it has something useful for or as a service to others, then you feel happy. Your life can be something useful. And so I've found my life to be meaningful, day in and day out, month after month and year after year. Wealthy people with a luxurious way of life may not experience satisfaction; they always want more and more. Without thinking, without helping other people, living an individualistic life—life becomes meaningless.

"We humans have this wonderful intelligence. We should use this intelligence to increase the world's happiness, to create peace, to bring greater compassion to society—sometimes I feel that's our destination. To try to make a contribution to developing a compassionate society."

"Icelanders have just engaged in a social experiment, a

human experiment, on this island of ours. We tried to see what would happen if everyone thought only about their own self-interest. It did not go well."

"Selfishness, you see, is our nature. If you don't take care of yourself, who does take care of you? It's very natural, reasonable, and realistic, thinking about yourself. But taking care of oneself at the expense of another's interests, that's wrong. Everybody has the right to live a happy life. We are social animals. Our attitude makes an impact on each other's happiness. Therefore, since we have the potential to express affection or compassion or a sense of concern for others, why not? In childhood, you see little intelligence. Day by day, year by year as you study, it develops. We strengthen and nurture the child's intelligence, but why not the child's love and warm-heartedness? That can develop, too, day by day, month after month, if we pay attention."

"You have not spoken about the Chinese with any bitterness in this interview. However, Tibet has been subjected to harmful treatment by them. Can forgiveness replace justice or punishment?"

"That's no contradiction. Forgiveness means you do not have hatred and anger toward those who did you wrong. Forgiveness does not mean that you accept others' injustices. We struggle against their unjust attitudes and policies, but we are all the same human beings. We have a rich, shared cultural heritage. The Chinese say that old Tibet was very bad, that the new Tibet under the regime of Chinese communists is much happier. We have to take this seriously, to research it. But as far as I remember, when we were in Tibet, before 1950, there were a hundred people in prison, at most. Now in the last fifty–sixty years, the prisoners number around seven thousand. Everywhere there are prisons, there is a lack of education. So I think you need to have education. Tibetans were once a jovial people. They lacked material infrastructure but were a very happy people. Now that joy's gone."

A monk appears in the doorway; our time is up. I say my farewell to His Holiness and thank him but he turns to me and asks:

"Have you ever been to India?"

"No, never."

"If you have the opportunity to come, let me know. We can discuss things further."

His Holiness the fourteenth Dalai Lama was generous and witty, warm, sorrowful, serious, and contemplative. He did not speak out against China in anger but instead talked about how Tibet had thrived inside China. Anyone who has been affected by war and exiled and still does not speak ill of those who have wronged him, that person shows remarkable strength. It is in our nature to be warriors, to take revenge; the Dalai Lama has the power to command violence but chooses not to.

The interview slipped past like any other big event in one's life: confirmation, graduation, a premiere. Your mind gets a little bit fogged up and you find yourself hyperaware of time flying by and then you experience a kind of release of tension. I read back over the interview and pondered its message. The importance of compassion, of cultivating the heart's tenderness. The purpose of life and of forgiveness. But what resonated were these words:

. . . in the next twenty to thirty years, some major rivers will be reduced, and some rivers may dry up. Before that, because of more melting of snow, there will be more floods. But then eventually it will dry up. So very serious. Usually I feel, and also tell people, that unless we take special care, if the present tendency continues, then I think in the next generation a billion human beings will suffer that threat.

What started out as an interesting opportunity to meet a truly remarkable person and some informal interviews with my aging relatives has suddenly begun to revolve around the most serious issues facing the world's population. Glacial melting can have the most drastic consequences; millions, even billions of people are at risk. Even if only 1 percent of those are affected, that's still ten million.

And the evidence mounts up. A 2019 comprehensive report, the *Hindu Kush Himalaya Assessment,* analyzes climate change's effects on the Himalayas and on the Hindu Kush mountains, suggesting that 30 percent of the glaciers will disappear by the end of the century and that even if humanity achieves the United Nations goals, stifling rising carbon dioxide emissions and keeping global warming within 1.5 degrees Celsius, that will not save those glaciers.[19] The glaciers have already begun to retreat at unprecedented speeds and although a 1.5-degree rise in Earth's temperature sounds small, it is still a prescription for even faster melting. The report does not mention one billion people but rather one and a half or two billion. Kunda Dixit described the report as "terrifying" in the *Nepali Times.*[20] If nothing is done to reduce greenhouse gas emissions and to prevent global warming from approaching the four-degree Celsius rise it is now heading toward, up to two-thirds of the glaciers could melt, with disastrous consequences.

After my interview with the Dalai Lama, a monk came to us and said that when His Holiness invited me so graciously to discuss matters further in India, he was entirely serious; we could meet him in a year's time at his home. And that was that: I was given time for another conversation, in Dharamsala, at the foot of the Himalayas, just under Audhumla's tail.

A revelation from the wrong god

Now the word of the LORD came unto Jonah the
son of Amittai, saying, "Arise, go to Nineveh,
that great city, and cry out against it; for their
wickedness has come up before Me."

—Jonah 1: 1–2

Humans have always been believers. We've stared in stunned wonder up into the starry sky. We've worshipped waterfalls and our forefathers, carved wooden figures, gods in the sky and from the depths of the sea. Mankind has had a single god, a hundred gods, demigods; we've had supernatural beings, goddesses, devils, demons, guardian angels, holy trees, and sacred sites. Now we have the big religions, Christianity, Judaism, Islam, Buddhism, and Hinduism, all of them divided into smaller sects and congregations.

When I was younger, I sometimes wondered how likely it was that the society I was born into had stumbled upon the right arrangement of divine forces, the sacred truth, the right god, an accurate idea of life after death. I wondered if the gods had perhaps split up the world into territories, and you might get a revelation from the wrong god, kind of like when someone phones the wrong number. I had an idea for a story where a mysterious elephant god appears to a young man who has no idea what god he is talking to. He hasn't a clue what this god stands for or what superpower he has.

But then Audhumla the universe cow appeared before me with an important message to the world. I'd called Gudmundur Páll, the person I knew who best understood Iceland's nature, and the world's, and I'd found he had the same story to tell. Audhumla had spoken to him, too. But right as we believe we've found Audhumla, all the research out there indicates her power is fading, that she might even be dying.

The Himalayan glaciers cumulate winter storms and monsoon rains and release them when people need water to persist through drought. The glaciers absorb seasonal fluctuations, but if they do not survive, exaggerated weather conditions will alternate between mass floods and droughts. In ancient Hindu writings, it is said that in the early days Ganges fell from the sky with enough force to destroy everything in his path. But Shiva saw this and caused Ganges to fall on his head so that the water seeped through his hair and flowed gently away to the people. The glaciers work this way: growing out from the mountains like Shiva's hair, they cling on to excessive, damaging waters and distribute them evenly throughout the year, for everyone's benefit. And it is not only agriculture that gains from this. The glacial rivers transport dissolved matter to the sea, which nourishes the ocean and algae and fish larvae; they live where fresh water meets salt water.

Earth's most densely populated regions lie around the Himalayas. Three nuclear powers surround them: Pakistan, India, and China. Pondering Audhumla, one can see why the Chinese covet Tibet and do not want to let it go. If they have Tibet, they have Audhumla, and whoever has Audhumla has control over Asia's principal water supply.

In my travels, I meet up with Satya Dam, an Indian naval officer and a great mountaineer. He is very worried.

"In the future, glacial melting may become a cause for war. What will people do when faced with water shortages in China, and supposing the government decides to channel Brahmaputra to China instead of allowing the water to flow to Bangladesh? What happens if droughts result in a dissolution of the historic Indus Waters Treaty and the exploitation of the rivers between India and Pakistan? It is obvious that glacial collapse is causing instability, crop failures, famine, conflicts, and even more serious disasters than we've previously seen."

The picture of the future he paints is alarming.

I met the Hindu guru Swami Nikailananda Saraswati and asked him about holy cows. He explains to me the cow's usefulness. The cow is the one who always gives. She gives milk and she produces calves; we use cows for agriculture and for milk, and from milk we can make ghee, which we use for lighting lamps and worshipping gods. With the help of cows, we can not only achieve material prosperity but also engage in spiritual rituals. The cow is the mother of both materialism and spirituality. Swami told me this story:

Once, long ago, there was a great disturbance on Earth. The demons started making a great ruckus. So Mother Earth took the form of a cow and she approached the Devatas, all the gods, and she said:

"Please protect me; the demons are troubling me, they are hurting me. Protect me."

The gods approached their god, the Brahma, the

creator. They said:

"Oh Brahma, the demons are hurting us. Hurting this Earth."

Brahma said, "I cannot do anything."

So the gods approached Lord Shiva and together they prayed to the almighty god who then manifested as a human being. He killed this demon and protected the Earth. And so we consider Earth to be a cow and represent her that way.

There was a time when I would have found this story bizarre, but this ancient Hindu narrative seems to have taken on a very literal meaning: the Earth appeared to me as a cow, begging me for help, on its own behalf and that of all its dependents.

What to do when the World Cow appears to you? How can I explain it? Should I tell the world one billion lives are at risk?

On whose behalf are you speaking? someone might ask.

I had a vision of the holy cow, Audhumla. I speak on her behalf.

Climate scientists have been accused of being harbingers of the apocalypse; in reality, they've been far too cautious. The course of events has exceeded their darkest forecasts. Many scientists refrain from making too much of their findings, afraid they will be accused of stoking mass hysteria. They also fear becoming the poster child for those who get it wrong. An end-of-the-world article in *New York* magazine described a devastating, horrific image of a world which will be "worse than you think"; the author warns us: "you are surely not alarmed enough."[21] In this future, the Earth has been plunged into total climatic chaos, with all nature's systems becoming unbalanced one after another. "Fleeing the coastline will not be enough," the author writes. The permafrost will have melted and methane gas will have flowed unfettered into the atmosphere. Greenland's glaciers will be teetering toward uncontrolled melting. The article's author describes a world of weather extremes, floods and famines, "perpetual war," a world with no way to sow or reap once all the ecosystems were out of whack.

I read it and thought: I must change the way I'm living, right now, entirely.

But then I read a discussion on Twitter in which leading climate scientists criticized the article. They pointed out that its conclusions were not 100 percent certain. Sure, the permafrost might melt but that idea required further research. Nor was it guaranteed that hundreds of millions would die; at most, it would be thirty million. One climate scientist criticized the journalist who'd painted this bleak picture, saying, "If you scare people, you'll paralyze them." Then a psychology professor entered the debate and asked the climate scientist to point out studies that showed how people facing overwhelming tasks entered paralysis. The climate scientist couldn't respond adequately; the psychologist hammered at him, asking: "When climate scientists soft-pedal their results, could it be that they're being armchair psychologists, acting out of an imagined fear of human paralysis?"

The psychologist pointed out that even when people react to terminal medical diagnoses, they aren't necessarily mentally paralyzed. Those who are physically paralyzed can respond with remarkable, sometimes astonishing vitality. He pointed out that humankind has survived all kinds of local apocalypses over the past millennia; people endure and more often than not overcome such problems, even emerging stronger. He described humankind as "antifragile": challenges, randomness, disorder do not bring us to our knees but, on the contrary, make us stronger; without challenge we wither and become weak.

Indeed, when humans face great threats we tend to show what we're made of. Most of human progress stems from having had to overcome a threat or an insurmountable obstacle: hunger, cold, wild beasts, climate change, gravity. Some of the greatest technological advancements of the twentieth century derive from World War I and World War II, when global nations feared imminent attack and labored toward startling advances in flight, radar technology, shelf life, telecommunications, and medicine, as well as unforeseen "progress" in murder and destruction. I do not like to think of war as a metaphor for technological advancement; the race to get to the moon is perhaps a better example, comparable in terms of progress achieved, but without direct suffering or bloodshed.

My thoughts turned to those who'd most inspired me in recent years. Gudmundur Páll Ólafsson labored to finish his book *Water in Icelandic Nature* while battling cancer, knowing he only had a few months to live. His book heeds Audhumla's call. It is an extended confession of love for Earth and for water on the Earth and a warning about what will happen if we do not turn aside from destruction. Why should he think about the future, knowing his life was soon to end? Do we have any purpose or role to play following our time on Earth? The Icelandic actress and director Edda Heidrún Backman, who founded the environmental organization Nature's Voice, was paralyzed from the neck down because of ALS (Lou Gehrig's Disease) and died as a result of it in October 2016. She had sent me a message in April that year, saying that it was time to assemble people to save the Earth. She gave me a painting of the mountain Herdubreid, which she'd painted with her mouth because all her other body parts are paralyzed. I distinctly remember what she said: "to save the Earth." I've always been embarrassed to use that verb, "save"; it feels dramatic, exaggerated. But the conclusion of the 2019 United Nations Intergovernmental Panel on Climate Change report states it very clearly. We are the last generation able to save the earth from irreversible destruction.[22]

The fact that Audhumla spoke to me is clearly an extreme example, but I think that most thoughtful people have felt at some point that it is high time they stir themselves to action and help save Earth—but then they find an excuse to avoid the hassle. My escape was fiction; I chose to let others worry about activism. I organized a massive book launch for my sci-fi novel *LoveStar* at the historic St. Mark's Bookshop in Manhattan. It was at 7:00 P.M., October 29, 2012. No one came to the party except Hurricane Sandy, which hit New York exactly that evening, at seven o'clock.

I have four children who are starting to make their own life choices. What should I tell them, how should I explain what's happening? I feel bad taking away their sense of purpose, their faith in the future. It's haunting to see their eyes dim as they read newspaper articles or watch YouTube videos about the Earth's decline over the next hundred years.

I feel a pang in my belly when they ask, "Has Earth been destroyed for us?"

Back in time

And time is like an image
painted half by water—
and half of it by me.
—Steinn Steinarr

Scientists talk about time: they debate and extrapolate to create a picture of the future. They run complex models through supercomputers and get answers to what the world might look like in 2050, 2070, 2090. We find it difficult to connect with and respond to such years. It's 2020 now, one year after the movie *Blade Runner* takes place. Five years after the future in *Back to the Future*. Thirty-six years after the Orwellian year, *1984*. We are so hypnotized by progress and revolutions that our relationship with the future is characterized by irresponsibility. For us, a hundred years is like a whole eternity, a thing beyond imagination. One hundred years seems such a long time that we don't react when a scientist shows us that if things keep going at this rate, substantial disasters will have occurred by 2100. We shrug our shoulders, as if that date doesn't concern us.

Scientists believe that human impacts have become so widespread that we are entering a new geological era. This modern era, the Anthropocene, branches off from the Holocene, which began ten thousand years ago. Since World War II, human activity has grown exponentially, whether in terms of population, consumption, energy use, or pollution. My Grandpa Björn and his second wife Peggy belong to a generation that was born and came of age at one geological time and now live in their old age at the dawn of a new geological age. When they were born, around 1920, the world's population was about 1.9 billion. Today, it's over seven billion. We are looking at nine billion by 2030 and more than ten billion by 2050.

I'm fortunate to have a direct connection to the past; I can ask my grandparents in person: "Is a hundred years a long or short time?"

Grandpa Björn was born in Bíldudalur in 1921; he died in 2019, when he was ninety-eight. I have often thought of writing his biography and even writing about his siblings and cousins. But his story fits in well with this narrative of time and water.

I met up with him in April 2018. He'd moved back to New Jersey from Florida, where he'd been planning to spend his retirement. He and Peggy had owned an apartment on Cape Canaveral, from the balcony of which they could watch the flares from NASA space launches. Then for two years in a row a large portion of Florida's population was forced to evacuate due to impending hurricanes, Matthew in 2016 and Irma a year later. Their daughters, my mom's half sisters, had to fly against the storm to get them to shelter; this clearly couldn't continue. Grandpa Björn gave up on Florida and returned to New Jersey.

"The hurricanes used to come maybe every thirty years," he says. "Then they started coming every year and that didn't work out so well."

"Is it because of climate change?" I ask.

"No one knows," says Grandpa. "Perhaps the pattern won't be visible until a thousand years have passed."

In becoming a climate refugee from Florida at his advanced age, Grandpa Björn was among the first but will be far from the last. Earth's sea level has been unusually stable for 2,500 years after rising 120 meters following the end of the Ice Age. But now humankind has started a new cycle of rising; according to scientists' forecasts, significant parts of Florida will be submerged in the life span of someone born now who gets to be as old as Grandpa.[23]

"How is it, being here?" I ask Grandpa Björn. They'd moved into a snug apartment in a picturesque seniors' home. He'd never intended to move to a seniors' home; after his feet gave out on him, he's been discontented with his loss of liberty. The authorities in New Jersey won't acknowledge his driver's license even though it was accepted as valid in Florida.

"It's a prison; I'm no longer allowed to drive."

"Don't be that way," Peggy says. "It's lovely here." Peggy looks great. She's always been cheerful, funny, and gregarious.

She has a strange electric cat in her arms that mews and purrs.

"Is the cat weird?" she says.

"Yes," I say.

"His name is Eddie," she says, "isn't he lovely?"

I pat the mechanical cat, which purrs oddly in response.

Lisa, her daughter, my mom's half sister, apologizes for the cat.

"Someone told me a robotic cat was good for old people, but I worry it's made Mom more eccentric. But you should have seen the retired surgeon replacing the cat's battery." Peggy is a great humorist. Much less guarded than my grandpa; I can't tell if she's kidding or not as she pets the robocat.

Grandpa Björn grew up in Bíldudalur, a fishing village of two hundred people in Iceland's remote western fjords; his father, Thorbjörn Thórdarson, was a district physician. Björn graduated from Akureyri High School in 1940, then went to medical school. Options were few and far between at the time. War was raging in Europe and anyone in Iceland who wanted to get an education had the choice of becoming a priest, doctor, lawyer, or teacher. He met Grandma Hulda at a ball in Reykjavík during the war but then went to America in 1949 for medical school when Mom and her twin sister were just three years old. He went for a year and a year turned into seventy years. My grandparents broke up in the mid-1950s; he settled in America for good.

Grandpa Björn advanced swiftly in his career, becoming a professor at Cornell University and a chief surgeon at New York Hospital, later New York-Presbyterian. He met Peggy, a nurse from Toronto, moved to the well-to-do New Jersey suburb of Norwood, and had four children. Björn drove to work in Manhattan each day. Grandma Hulda meantime found herself a mountain man back in Iceland and had two more children.

Grandpa Björn's face is etched with age, he's unsteady

on his feet and hard of hearing, but his mind is a steel trap. We alternate between Icelandic and English, our conversation flowing across languages. He's long had a hard time hearing but recent advances in hearing aids have made conversation easier. I ask him about time, now that he's the oldest man I know. How is it to be almost a hundred years old?

"*Jæja*," he says. "Everyone is dead, almost all my co-workers, all my old neighbors and classmates and all my siblings are dead. But otherwise I'm just fine. Of all my school friends from Akureyri High School, there's probably no one left alive, except Önundur Ásgeirsson. Is he alive?"

"I'll check," I say, picking up my phone and googling the name. Grandpa Björn follows along. His eyes squint to tiny slits in his face. The phone loads an obituary from *Morgunbladid*.

"Önundur apparently died in February," I say.

"Oh well," says Grandpa Björn, heaving a sigh. "Then I'm probably all alone." At that moment, he was the oldest living Icelandic doctor, the oldest remaining graduate from his high school. I ask if he feels like it's been a long time, his lifetime. He says that a hundred years is no time at all. "I feel like it was just yesterday that I was working on a herring boat, the *Westan*, in Arnarfjördur."

Grandpa Björn was in the world news when he operated on Mohammad Reza Pahlavi, the Shah of Iran, at the end of October 1979. I was six at the time and living in New Hampshire. I heard Grandpa's name on the radio as we drove to my classmate's birthday party at Burger King. I asked my grandpa about the operation and he told me that the Shah had been downhearted. As he was being admitted, he heard from the news on the radio about close friends and colleagues whom revolutionaries had brought out and executed. Grandpa Björn told me about the crowds outside the hospital protesting against the Shah. Or, as he put it in an article about the Shah's spleen: "Outside, the hospital was surrounded by a howling mob, controlled by barricades, calling for the Shah's head."[24] Grandpa Björn's daughters were fifteen and twenty; the CIA instructed them to take precautions if they saw any mysterious people following them. "What should we do?" they asked. "Run."

Insurgents in Iran fervently protested the Shah receiving medical care in the United States. Four days after the operation, on November 4, 1979, the U.S. Embassy in Tehran was taken hostage by a group of fundamentalist students. They demanded the Shah be extradited to Iran and the embassy staff were held hostage for 444 days. An unsuccessful rescue operation ended the Carter administration and Ronald Reagan took over. The Shah was stateless; no one seemed willing to accept him. He traveled to Texas and Panama and eventually died in Egypt in the summer of 1980. The Shah had been a key figure in defending U.S. oil interests in Iran. An oil crisis followed the revolution: oil production in Iran reduced from 6 million barrels (950,000 m^3) per day to about 1.5 million barrels (240,000 m^3) and world crude oil prices doubled amid global economic turmoil.

"Did you find it difficult, being at the center of world attention?"

"No, no. I treated all my patients equally."

As a New York chief surgeon, he received patients from

all over the world. A sultan came flying in on a jumbo jet, his retinue in another, because he'd found out he had tapeworms.

Andy Warhol was Grandpa's patient in 1987.

"Warhol was an odd fellow," says Grandpa. "He'd gotten a phobia of hospitals after that woman had shot him several years earlier. He comes to my office and tells me, 'You skip the operation and I'll make you a very rich man.' He'd hidden his illness for weeks and was in critical condition. When I operated, his gallbladder was gangrenous, but the operation was otherwise a success. Warhol was in high spirits when I said goodbye to him, but then I got a call that night. He'd died."

This was, as one might imagine, a great shock; the media said it had been a minor, "routine" operation. This isn't the fifteen minutes of fame surgeons want. When I met up with Grandpa Björn in April 2019, the *New York Times* had just published a major article about the operation, saying that it was not a minor procedure but a major operation on a frail patient and that no one should have been surprised by the risks involved.[25]

Grandpa Björn also operated on Robert Oppenheimer, probably the figure from the twentieth century who comes closest to a mythological existence. In Greek mythology, Prometheus brought fire to humanity by stealing it from the highest peak of Mount Olympus; Oppenheimer dove down into the smallest unit of matter and brought the world's leaders the nuclear bomb. As a result, the world's leaders had godly superpowers, the ability to blow up the whole Earth. Fallible leaders had been given the power of annihilation: a force multiple times more destructive than that of all the combined tyrants in human history. By comparison, Prometheus had a dim, tiny flare.

Many scientists have argued that Oppenheimer's work marked a new era in geological history. Radiation from the Trinity test established a layer we might understand as the starting point of the Anthropocene. As of July 16, 1945, human impact, our footprint, was formally measurable on all surfaces of the Earth, in all soils, all stones and metals. Humanity as an organism had begun to affect Earth the way major geological phenomena do. A single human now had the power to destroy all life with one decision: detonating a nuclear bomb. At the same time, other human activities have become so extensive that the years since World War II have been characterized as the "Great Acceleration." During this time, all humanity's effects on the Earth have grown exponentially; biodiversity has gone the other way.

Oppenheimer was more influential than any shah, president, or commander; he was of a mythical scale and his future legacy will depend on how we handle nuclear energy. Some narratives construe him as having brought peace to humanity, but the future is unwritten, and his influence can best be measured in legendary terms: The nuclear bomb hangs over us like the sword of Damocles. The peace it brought us might be a Pyrrhic victory.

The gods were angry with Prometheus and he was punished cruelly, tied to a rock where an eagle ripped out his liver. I say to Grandpa Björn, whose name means "bear": "I'm writing a mythology for the contemporary era. If an eagle gnawed Prometheus's liver out, what did the bear do to Oppenheimer?"

Grandpa Björn thinks a while and smiles a little.

"I signed the Hippocratic oath; unfortunately, I can't tell you."

"Was it something to do with the liver?" I ask.

"Just say it was hemorrhoids."

Time shrinks into the distance. In five thousand years, Oppenheimer and Prometheus will have practically become contemporaries in the eyes of the people living then. The paper used in most modern books only lasts a hundred years. If I write this story with bearberry ink on calfskin and sneak it into the Árni Magnússon Institute's archives, it might be preserved for a thousand years, perhaps more. The mythology the future receives might sound something like this: An eagle gnawed Prometheus's liver while a bear treated Oppenheimer's hemorrhoids.

Grandpa Björn tells me that as Oppenheimer was lying on the operating table, an image flashed before his mind's eye, from within his childhood church at Bíldudalur. Oppenheimer was strikingly similar to the image of Jesus Christ on the old altarpiece there. "It's not so odd," Grandpa says. "They were both Jews and both in their own way revolutionized the world."

Oppenheimer himself realized the mythical context of his actions when he saw the bomb explode for the first time. He said in an interview:

We knew the world would not be the same. A few people laughed, a few people cried. Most people were silent. I remembered a line from Hindu scripture, the

Bhagavad Gita: *Now I am become Death, the destroyer of worlds.*[26]

My generation had recurring nightmares about Oppenheimer's discovery. Were our worries unnecessary? Or were the worries the reason the world had survived? I was brought up with photos of Hiroshima, and the terms "nuclear winter" and "nuclear fallout" hung over my childhood like dark clouds. These words were charged, fully so; there was no buzz in them, no white noise.

And now scientists have drawn up new horror scenarios about the consequences of a three-to-five-degree rise in Earth's temperature. The Earth has abandoned geological speed; it is changing at human speed. And yet our response happens at a glacial pace. We hold a conference to determine the location for the next conference. Perhaps this increase does not contain enough of a sense of urgency, a fast-enough rate of change, not compared to sudden flashes and shock waves. We look quite calmly up at a slowly developing disaster. "Global warming" sounds totally different from "nuclear winter." A few more forests burn here, a little more heat there, things even improve a bit for a little while—until suddenly a thousand-year flood rushes down on the calmly rising sea level, with slowly expanding deserts and slightly stronger hurricanes. All the while, the dying out of one species or another ceases to be newsworthy.

My generation doesn't seem to want a return to fear. But, unfortunately, we have to take this seriously. Our children have not experienced nuclear threat or the hole in the ozone layer; they have not manifested an ironic attitude toward apocalypse. They aren't afraid of being afraid. They're on strike. School ought to prepare them for their future, of course it should, but all education is meaningless if the school system and business communities don't adapt to science, instead leading us in the wrong direction. Over the edge.

Crocodile dreams

Grandpa Björn lived with Peggy and their four children in a big white house in Norwood, New Jersey. When I was a child and living in America, we often visited them at their two-story, fairy-tale house with its backyard swimming pool. They had a dog, a guinea pig, a dwarf caiman crocodile, and lots of mice to feed to BC, the nickname for their three-meter-long boa constrictor. Their eldest son, John Thorbjarnarson, was the owner of the crocodile and the boa.

One of the most enjoyable days of my childhood is captured on an 8 mm film my mom took. I was six or seven and we were visiting New Jersey when John came outside with the snake. We got to hold it and he allowed it to swim with us in the pool. That same evening, we went night swimming while bats caught the moths that were seeking out the glowing lights at the bottom of the pool.

Uncle John, who was born in 1957, had an insatiable interest in reptiles and amphibians: he loved frogs, turtles, and snakes he found in the woods behind his house. When John was ten years old, he watched a National Geographic show about crocodiles and the dangers they faced in the Florida Everglades. He seemed to have found his calling, an idea that when he grew up he could somehow work researching and rescuing crocodiles. And he went on to eventually become a herpetologist, specializing in the conservation of crocodiles.

Crocodiles have lived on Earth for more than seventy million years without ever having been much different from the way we know them today. Their evolutionary history can be traced back two hundred million years. They have adapted to tectonic movements and ice ages. They survived when the dinosaurs and almost three-quarters of Earth's entire biosphere died out in the fifth and most recent great extinction, sixty-six million years ago, when scientists believe an asteroid struck Earth.

At the time John completed his doctoral work in herpetology, the majority of the twenty-three species of crocodile that exist were on the shortlist for, or considered close to, acute extinction risk. His thesis dealt with the spectacled caiman, the same species of crocodile he'd had as a child. He subsequently collaborated on a comprehensive action plan for protecting crocodiles worldwide.[27]

I'd long dreamed of living awhile in the Amazon with John, writing a book or a series of articles about the river, the rainforest, and about the local culture. The first work of mine to appear in print was a translation of one of John's articles on crocodiles, turtles, and anacondas for the Sunday edition of *Morgunbladid* newspaper.[28] Translating an article about non-native species into Icelandic has its challenges. Our language doesn't distinguish between crocodile and alligator, never mind the caiman species the article was about—the black caiman is South America's largest predator. It can grow to more than five meters long and weigh more than a thousand kilograms.

We have seventy words for snow in Icelandic but only the word *krókódíll* for alligator, caiman, gharial, and crocodile. Imagine using only "sheep" to name sheep, goats, and antelope, and only "cod" for cod, coalfish, salmon, and haddock.

In 2010, John published a book in which he argued that the Chinese alligator was the inspiration for the Chinese dragon. When his colleagues began counting this species, only about 150 adult alligators remained in the wild, living in small ponds and backyards in Anhui province on the Yangtze River.[29]

For seventy million years, crocodiles have survived ice ages and meteor showers. If you were to plot crocodile stocks on a million-year-long x-axis, their population would fluctuate; every million years or so a particular species might die out from natural causes—but in the twentieth century there's a vivid vertical decline, as with most animal species. Humans have encroached on habitats and feeding grounds. We covet boots and bags. We are ignorant of and hostile toward our "ugliest" cousins. We live in an era where a few people can create a bomb that equals a meteorite storm and a simple fashion trend can become a pestilence for plant and animal life.

Scientists have pointed out that we are experiencing the sixth mass extinction period for animal species in the Earth's history. The five preceding extinctions are:

1. 444 million years ago, at the end of the Ordovician period.
2. 375 million years ago, at the end of the Devonian era.
3. 251 million years ago, at the end of the Permian period.
4. 201 million years ago, at the end of the Triassic period.
5. 66 million years ago, at the boundary between the Cretaceous and Paleogene periods.

The sixth extinction began with the modern era, the Holocene, the Age of Humanity that commenced 11,000 years ago with the eradication of megafauna from the continents. But all this has escalated in recent decades. Our consumer habits are volcanic in their effect, our fashion trends have more impact than tectonic shifts, our desires are earthquakes. We live in a time when the initiative of a few individuals can influence whether the strongest, most powerful predators on Earth—our era's dragons—will live or die.

Despite having been born in a New Jersey suburb, John, together with his colleagues at the Wildlife Conservation Society, has influenced the sixty-million-year evolutionary history of the species and got in the way of its eventual destruction. You need a particular personality for such work. Crocodiles are usually found in wetlands and marshlands, more often than not in remote, impoverished regions in less developed countries, where any type of government might wield power. John needed to get locals on his side, to understand their position and the struggles they face while at the same time reinforcing the role these predators play in nature, how predators are more often than not a vital part of a healthy ecosystem. If they live as part of the ecosystem, they don't take anything away from nature; they cycle nutrients and raw materials, and maintain habitats. Crocodiles catch weaker animals and prevent overpopulation and disease. They dig holes in rivers and marshes to prepare for drought periods; these holes benefit other species and plants.

John was quiet and modest and he had a subtle wit, a long way from the brash and macho types who wrestle crocodiles on television. He respected animals, countering people's prejudices by talking about crocodiles as gentle creatures. He enthused about how there's nothing more beautiful than canoeing in the middle of the night in the Mamirauá Reserve, flashing a light into the swampland and seeing a thousand gleaming eyes in the dark. He once told me crocodiles actually behave more like birds than snakes and other reptiles. They make nests and lay eggs in them. Spectacled caimans hear the call of their young when they're ready to crawl out of the egg. The mother hustles to the nest to protect them. She gently takes the eggs in her mouth, showing maternal warmth as she helps them hatch one by one, gently rolling them between her tongue and gums. Once the young have crawled out of the eggs, the mother ferries her offspring in her enormous jaws, placing them in a safe spot. Sometimes she has to make several trips. Then she attends to her young crowd like an eider duck until they become self-reliant.

Although John worked to conserve endangered species, I don't remember him ever being angry or accusatory. I don't recall him campaigning on the Internet; the kind of people he had to approach would have been unlikely to notice, let alone engage in, such provocation. He worked with farmers and hunters to plan stocks and conserve habitats and nesting places. He negotiated with wealthy landowners, poor farmers, and bureaucrats. It isn't always economic arguments that matter most or even the philosophical argument that animals have a right to their existence. Sometimes direct, clear arguments about beauty matter most, the awesome respect for the species or its history; its role in folklore and musical heritage; or else man's ability to admire, to fear and be amazed by it, to perceive in a wild beast some immeasurable depth of inner life. The crocodile's mythical dimension, in other words. The beast that gave rise to the idea of the Chinese dragon and the Egyptian water god, who made the grass green.

Of course, humankind has affected animal species in our struggle for bread, toiling against hunger and disease. The problem is, humans don't seem to know any boundaries. They do not know when they're satisfied, nor when they've gone too far. Expanding one's rice fields to feed a large family is not the same as filling in a pond, eradicating its fish and flowers and nesting sites, in order to build a shopping mall or amusement park; not the same as clearing rainforest in Jamaica and drowning mountain valleys in Iceland to produce cans of soda. In Iceland, there are disputes about the Svartá hydroelectric power plant, where there are habitats for falcons and Barrow's goldeneye ducks, where you find quacking harlequin ducks and sea trout. They want to disturb it all for a 10-megawatt power plant. To put this in perspective, Icelanders use 150 megawatts mining Bitcoin, virtual currency with no evident purpose. People gorge on a twelve-course buffet, then tickle their throats with a feather to make themselves throw up so they've room for more, for extra helpings of cake. In the Westfjords, a precious highland plateau is being disrupted for another 50 megawatts. It's as if people think that disturbance is the goal, a purpose in and of itself, as if they're asking: "Why is this area here? What's the benefit of an area nobody uses?"

Laozi's *Tao Te Ching*, or the *Classic of the Way and of Virtue*, was written in China some time between the fourth and sixth centuries BCE and published in Icelandic around 2,500 years later, in 1921, with the title *Bókin um veginn*, or *The Book of the Way*. My favorite section is about usefulness:

> Thirty spokes coming together make up a wheel
> but it is the hole for the axle that makes the wheel useful.

> We throw clay to shape pots
> which work because they are hollow.

> Men cut out doors and windows
> and the house's empty space inside makes it useful.

> For existence to bear fruit
> what does not exist is most useful.[30]

Emptiness allows the wheel to revolve. Throughout the twentieth century, we have demanded that the Earth turn a profit, that it produce ever greater output. We have filled up more and more of its emptiness. We've called that common sense. What's this marshland for? Why all these flies?

Can't we get rid of the competition from foxes and crocodiles? For an area to be protected, it needs to serve as a national park or tourist destination; its purpose needs to be measurable, preferably by how the national park directly delivers dividends, employment, and increased sales of goods and services in close proximity. When people discuss coral reefs, they talk about their importance for fishing and tourism. Art needs to justify itself with sales and turnover. Education and science need justifying on the basis of the products and the employment they create. What does not exist is not considered useful in itself; emptiness, the axle's hole, is constantly impinged upon, until it now seems life's wheels have stopped turning.

Scientists have measured and estimated the decline of wild nature in almost all Earth's spaces, whether it be rhinos, guillemots, bats and puffins, orangutans, bears, fish stocks, or the reindeer in the Arctic tundra that died off in the tens of thousands in the winter of 2018–19.[31] Scientists have become aware of an unexpected global decline in insect populations. Truck drivers in the U.S. talk about windshields that used to get blacked out by flies now making it through entire states clean as a whistle. About 75 percent of flying insects seem to have disappeared, based on research undertaken in protected areas in Germany; this, of course, has repercussions for other life-forms. Even in unspoiled rainforests humans have gone nowhere near, there's been a decline in insect numbers.[32] The "sixth extinction" sounds dramatic and maybe we're so close to the black hole that we can't see it. The image is bleak, the soundscape all buzz, but the fact remains: we're facing a totally unprecedented task.

When John and his colleagues formed an international network of crocodile scientists and combined their strength, about twenty-one of the twenty-three crocodile species were on the endangered species list; after the actions of biologists, many species have recovered, or, at least, are beginning to. In 2010, only seven were still classified as endangered, according to a *New York Times* article about John's life work, an article that, sadly, was also his obituary.

On February 14, 2010, we received the sad news that John had died, aged just fifty-two.[33] He was in New Delhi, India, having arrived recently from Uganda, where he worked with locals to research rare dwarf varieties of Nile crocodile. He was in India to give a talk to those working to conserve the gharial, which has mostly disappeared from its natural river environment, where it has lived for millions of years. We knew John's job was dangerous, but he always said that crocodiles were not the main threat. More serious were traffic, food poisoning, and malaria. It was probably a resistant strain of malaria that killed him.

I never knew how well-respected John was; he never boasted about his work. The *Economist* dedicated a whole page to his obituary, as did the *New York Times*.[34] Thanks to him, the Orinoco crocodile and Chinese alligator have begun to recover, and work on restoring the black caiman and the Indian gharial is underway. Rescuing an entire animal species is probably one of the most noble things a person can do in a lifetime, but the fact that a small group of people can have a decisive effect on the history of a millions-of-years-old species shows how unique our time is. So many species are in danger today that some biologists feel they have been drafted on board Noah's Ark. All over the world, people work to save species in the hope that they will survive after humankind's contemporary bomb of consumption and waste comes to an end. Perhaps a new period of maturity is emerging, in which all Earth's species are respected and have the right to life and habitat.

The dying off of animal species through human agency is a major event. In Iceland we lost the great auk on June 3, 1844, when the last two birds were killed in Eldey, a small rocky island of the southwest coast of Iceland. The scientific name of the bird is *Pinguinus impennis* and it would have been nice to have our own Nordic penguin, but unfortunately people were not in the habit of protecting animals back then. In 1929, there was a debate about the great auk in Icelandic papers after an ornithologist, Peter Nielsen, wrote an article in *Morgunbladid* about the last great auks and their killing. He described how the specimens became more valuable to collectors the less common the bird became. Eventually, fishermen could earn a full season's salary catching birds and retrieving eggs for European collectors, which contributed to an even faster destruction of the species. Ólafur Ketilsson, the son of the man often accused of killing the last great auk, replied to Peter in a separate article, defending his father's honor. Peter answered Ólafur, reminding him that he had not mentioned any names in his article and explaining why:

Out of reverence for the memory of a dead man, and so as not to hurt the feelings of his relatives or friends, I did not mention by name the person reported to me

as having killed the last great auks known to exist anywhere in the world, or at least the person who had been
so unfortunate as to cause their deaths. I thought it
inappropriate for the public and unnecessary to air.[35]

But because Ólafur had come forward, Peter Nielsen took
the opportunity to emphasize what had happened that
fateful June day in 1844, citing a Danish report:

> If this report is correct, and there seems to me no reason
> to doubt it, this man who hunted the last-but-one auk
> was named Jón and Sigurdur was the one who hunted
> the very last—but the man who took hold of the last
> auk's egg, which in all probability was teeming with
> life—Professor Newton says that person was called
> Ketill. It is thus not at all improbable that it was Ketill
> who killed the last bird!

According to *Morgunbladid*, Peter Nielsen was born in 1844,
three months and five days before the death of the last
great auk, and was therefore not likely to have been fortunate enough to set eyes on this creature. In 1929, he was
eighty-six years old and had been paralyzed for nineteen
years, but this article demonstrates he was clearheaded and
quite incisive in his words. He points out the importance of
learning from the fate of this bird, given that the sea eagle
and the gyrfalcon were headed the same way. In 1929, only
a very few eagles were left in Iceland after years of hound-
ing and harrying. They were shot and farmers destroyed
their nests. Farmers also laid out poisoned carcasses to
kill foxes; eagles would eat the carcasses, and so farmers

killed off many, many birds that way, too, all in order to protect sheep and eider ducks. Just like Uncle John, Peter Nielsen found himself discussing an unpopular predator which caused farmers perceived or actual damage. In a 1919 article, he states:

> Nothing in nature is without purpose. The eagle absolutely has its role. Every encroachment on the work of creation can have unexpected consequences, though they can't be seen at first.[36]

Nielsen's warnings and those of his associates just over a hundred years ago prevented the eagle and the falcon from going extinct. However, there are still farmers who believe that eagles plunder eider duck nests and so the farmers persist in stealing the eagles' eggs, trying to force them away from their territory.

The problem today, unlike in 1919, is that habitual local conservation is effective only within ecosystems where some kind of consistency and equilibrium already exists— if equilibrium can really be applied to nature. Animals live in concert with the complex interplay of seasons, a rhythm of rainy periods and spring floods, of water temperature, of the blooming of buds, of the hatching of flies and of fry, or of the migration of fish. If the foundation itself fails, if weather systems collapse, if the average temperature rises or ocean acidity increases, there is little individuals can do. If puffins or guillemots are protected, they will still decline over a few decades if there is no food for them—until at last they're gone, just like the great auk. No point holding up one particular hunter as the culprit; the root cause lies elsewhere. All these attacks on the works of creation are beginning to have unpleasant consequences.

According to a 2019 United Nations Intergovernmental Science-Policy Platform on Biodiversity and Ecosystem Services Global Assessment, just under a million animal species are in danger of extinction. The report was put together by a hundred fifty scientists from fifty countries. Their conclusion means that, if we put together the rapidly increasing destruction of habitats, industrial and agricultural pollution, overfishing and overexploitation, plus the effects of climate change, we have a real chance of the total collapse of entire ecosystems. All these elements cut so close to the Earth, so manifestly impinge on the future of humankind itself, that the report's authors call for immediate action.[37]

Global warming can be seen as a total shift in nature, with the foundational conditions for entire ecosystems shifting simultaneously. The world's ecosystems are pushing away from the Equator toward the poles and up mountains, fleeing rising temperatures at speeds of greater than one meter a day: southern fish species have moved up to seventy kilometers over the past decade while tropical organisms are increasingly moving north, bringing tropical diseases with them. On the Icelandic coast we have seen the mackerel come up from the south and the capelin disappear. Certain animals and plants can move and have moved, but complex ecosystems cannot be shifting locations in a single human lifetime, even if individual species do.

To understand the significance of a two-degree rise for plants and animals, we can look closely at our own bodies. For a human being, life would be unbearable if you always had a temperature of thirty-nine degrees Celsius, just two degrees above the norm. This describes in great simplicity pretty much what would happen if the earth warms by two degrees. Species adapted to particular conditions would suddenly feel hot, tired, and weak. Some would die. Others expend all their energy defending themselves against the heat; as a result, they can stop proliferating. Two degrees Celsius would be the average temperature increase across earth; more local changes could even exceed six degrees, which would overthrow the very basis of the biota. Some species would migrate, but more often than not an ecosystem collapses like a house of cards once the timing of bird migration, of hatching and budding, and of flowering goes out of sync. Sometimes, animals are already at the ends of the world and have nowhere else to go. If a species' ideal region is Iceland's northern coast, it's easier said than done to move farther north, where there's nothing but rough seas.

The animals are Earth's fruit; they grow like apples on trees. If a tree withers, it bears no fruit. It is useless to protect or preserve the apples while the tree is being chopped down or its roots gnawed at. As Árni Einarsson, a specialist on the protection of Lake Mývatn, said: "I have spent my life defending the flora and fauna of Lake Mývatn by nurturing the birds and midges, but then someone comes and changes the climate itself. And so the risk is that everything was in vain."

When Grandpa Björn and Peggy were born, there were people alive who'd lived in the time of the great auk. Back then, the shame of the killing of the last auks was on the conscience of Ketill's descendants, and on the descendants of his companions. The hunters did not know for certain that those were the last birds. The world often makes more sense in hindsight than when people are in the middle of things. If we look at scientists' predictions and don't do something radical right now, similar judgments will be made by the future about this sixth extinction. Our existence will be shrouded in shame. Our entire story will likely be heavy with meaning because of the consequences. We knew what was happening. We were all Ketill.

A mythology for the present

"What are you writing about?" Mom asks. She's just finished playing golf with my dad. She's wearing a pink sleeveless sports top, although it's barely six degrees Celsius outside.

"I'm collecting stories," I reply, "about the honeymoon on Vatnajökull, about Grandpa Björn in America, and about crocodiles."

"You could at least mention you have parents," Mom says, not exactly angry, though I sense a passive-aggressive tone and feel a little guilty.

I find myself thinking: What is it with grandparents?

Arguably, mythology began with the worship of ancestors—by putting our ancestors on a pedestal, smoothing out the imperfections and difficulties that are a part of anyone and everyone. We have a whole profession of people to help us disentangle ourselves from our parent-child relationships; everything parents do becomes fodder for psychological analysis. Too much intimacy; too much coldness. A father or mother's absence constitutes a problem; losing a parent is a profound blow. But if you have grandparents, you are fortunate, plain and simple—if you don't have any, that's not seen as problematic. Their presence can bring you an infinite number of things, but if they aren't around, it's not treated as though something's missing. While our relationship with our parents can be full of knots, relationships with grandparents are usually simple. Grandparents become heroes and demigods in the minds of grandchildren, leaving their parents to grumble: "Well, you think she's great, but Mother wasn't perfect . . ."

Jón Pétursson, my paternal grandpa, lived a life that was entirely different from that of Björn, my maternal grandpa. He did not particularly pursue a career or chase after material gain, spending more time on social issues and ideals. When he was fifty, he rejected a permanent position at the Reykjavík Water Utility Board so that he and Grandma Dísa could spend four months of the year at the abandoned farm his ancestors had lived in, on Melrakkaslétta, on the coast of Iceland's northeast corner. There, they took care of the eider ducks, fished for trout, and lived their lives. He died in 2006 and I regret not having interviewed him more often, but back then my mind was always up in the highlands. Grandpa Jón was born fifteen years after the first car was shipped to Iceland, before radio came into the story. Everything the family ate was caught or farmed on their own land; they raised sheep, milked cows, fished for trout and seals. Sugar, flour, and coffee were almost the only things they bought for the household when he was a child.

One time we were sitting in the little kitchen at the house in Teigagerdi in Reykjavík and Grandpa Jón was telling me about sailing to Blackpool in England with supplies of fish during World War II. Their fishing boat was part of a shipping convoy under warship protection. The boat had to sail with all its lights off; they weren't to stop under any circumstances, as German submarines made constant attacks on ships. If something happened, they were to maintain full sail; it was the role of the warships to come to the rescue. One morning, they were traveling in calm weather and fog and sailed through the wreckage of a sunken ship. They saw a white spot in the sea and realized that they had sailed through a group of floating nurses. I was alarmed when he said this, but he did not explain it any further. He was often confused and frail at this time. I never found out if the nurses were alive or dead or how they got into the sea, whether a hospital ship had been sunk. I'm not sure if this was a dream or a reality but the image has always stayed with me. Sailing at full steam in fog through a patch of floating nurses. Maybe this sums up Grandpa Jón. He'd sometimes recite a famous poem and then ask: "Did I write that?" And then burst out in gales of laughter. But he'd never said anything else quite like the comment about the nurses. I decided to ask my dad if he knew this story, but he wasn't even sure that Grandpa Jón had really sailed during the war; my aunt wasn't sure either. Time passes, things are forgotten, and the next generation loses the memories if you don't get around to asking, writing, or recording.

At ninety-eight years old, Grandpa Björn could remember everything. I could ask him about a surgery in 1970 and he'd remember which doctor had referred the patient to him. His siblings also have remarkable stories. Arndís Thorbjarnardóttir, his sister, was born in 1910. When she was about twenty, she headed overseas to Oxford to look after the children of a young professor of medieval literature. He was unknown at the time, and was just beginning to write a book that he was calling *The Hobbit*, in order to entertain Christopher, his youngest son.

I did not spend time with Arndís often but around the turn of the millennium I visited her at Grund, a seniors' home in Reykjavík, when Grandpa Björn was visiting Iceland. She told me the Tolkien family life had been a bit dull. The wife did not seem at home in Oxford and felt a bit inferior amid this highly educated society. She had lots of dresses she never wore and a piano she never played. It annoyed her to have a young woman in her home who spoke some kind of Elvish to her husband; he wanted to learn to speak Icelandic and this annoyed the wife. Arndís told the boys bedtime stories about Icelandic customs, stories about monsters and trolls, about Iceland's elves or "hidden folk," about volcanoes and about life in turf houses. She told me that Tolkien asked her to keep the door open so he could sit out in the hallway and listen to the stories, but that did not go down well with the lady of the house. Arndís looked after Christopher, who was quite energetic, teaching him Icelandic games and verses; she told me that when she read *The Hobbit* a great deal of it was familiar to her.

I can imagine Tolkien sitting there in his library with a pipe, hearing chattering and songs from outside in the garden where Arndís and the children are dancing and singing:

> Í grænni lautu
> thar geymi ég hringinn,
> sem mér var gefinn

og hvar er hann nú?

Tolkien asks Arndís to translate the song and she says, "In the green hollow, I hid the ring that was given to me, but where is it now?"

Back home, Iceland was rising and a wave of optimism and progress swept across the land, despite the Great Depression. The thousandth anniversary of the parliament at Thingvellir would be commemorated at the summer solstice of 1930; preparations for the festival had been underway for five years. Kristján X, the King of Denmark, was going to open the celebration, and warships carrying Swedish and Norwegian royal heirs would be traveling to Iceland.

Arndís was living in the home where one of the twentieth century's greatest mythologies was being composed but she hastened home to Iceland because she thought she was missing out on something exciting. Staying in Oxford does sound rather like staying in a quiet Hobbit hole compared to this spectacular event at Thingvellir, which sounds like a scene straight out of *Lord of the Rings*: the elf maiden rushing home as the royal heirs flock to the holiest place on the volcanic island to laud the thousand-year parliament.

I never met their brother, Páll, but he was the captain aboard the *Skaftfellingur*, which had its port in the Westman Islands and sailed with a cargo of fish to Britain during the war years. The crew was constantly on alert, not least because the sea was littered with mines and German submarines and aircraft repeatedly committed mass murders of Icelandic sailors. *Skaftfellingur* was the ship that first came to the aid of the trawler *Frodo*, which had been shot apart in a brutal attack in

20 Northmoor Road

Oxford

1941 during which five sailors were killed. German airplanes even fired at unarmed sailors in rescue boats.

In August 1942, the crew of the *Skaftfellingur* became aware of a vague bump half submerged in the water ahead; it turned out to be a damaged German submarine. They could see the frantic crew hanging on top of the submarine's tower; the boat was sinking. The *Skaftfellingur* numbered seven crew, sharing one gun, yet they rescued fifty-two German submariners. Páll was interrogated by the British army when he delivered the Germans to them. The British sailors simply could not understand this daredevilry one bit, why hadn't they just let the submariners drown? Fifty-two strong, they would easily have been able to commandeer the fishing boat and sail it back to Germany. Sometimes I've found myself wondering if this was an act of kindness or whether it was simply that sailors from the Westman Islands would have had such dire experiences of death and drowning that the idea of doing nothing to keep young people from dying at sea was impossible to imagine or let happen.

The world is full of stories; far too many of them disappear into fog. It would take a whole life to gather a life of stories. I asked Grandpa Björn when he thought the most remarkable changes had taken place in the world since he was born.

Without hesitation, he said: "During the last ten years."

Grandpa Björn had lived a full life and seen a lot: he was born in Bíldudalur where he fished herring on primitive steamboats; as a young physician, he'd traveled around the Westfjords on horseback visiting patients; and, later, he jetted out into the world, plunging into the New York whirlpool, coming into contact with the Iranian revolution and finding Prometheus himself under his knife. Grandpa Björn had seen it all, but still he said "the last ten years." That means computers, the Internet, genetic science, social media, information technology, Skyping the grandchildren. And all of a sudden, I don't think I've missed anything. It would be ironic to lose oneself in old stories and at the same time forget to pay attention to your own contemporary moment—and especially the future.

If we look at measurements and data, Grandpa Björn is statistically correct. In a world where changes in many areas have been exponential, there have been more changes in the last decade than in the entire twentieth century. Around the year 2000, roughly fifty-eight million cars were produced annually in the world; now it is a hundred million a year. Half of all the plastic produced in the world came into existence after 2000.

In the last ten years, we've seen the eight hottest years since temperature record-keeping began in the mid-nineteenth century. Since the turn of the century, the Icelandic glaciers have retreated more than they retreated in the entire previous hundred years. There's a reason to take notice of the present. The time of greatest change is upon us.

N 64° 35.378′, W 16° 44.691′

Grandma Hulda and Grandpa Árni's wedding photo stands on a small table in the living room of the house at Hladbær. They're gazing toward the horizon, standing right in front of Hvannadalshnúkur, Iceland's highest peak, 2,119 meters high. It was the Icelandic Glaciological Society's fifth trip since its establishment five years before.

They married on Friday, May 25, 1956, and the very next day met a nine-automobile caravan, including three snow-mobiles and an energetic group of traveling companions. Gudmundur Jónasson, a mountain driver, and Dr. Sigurdur Thórarinsson, a geologist, had made a survey flight over Vatnajökull to assess whether the route was feasible; still, shifting glaciers and volcanic activity could easily derail the expedition. It was a large undertaking and they had enough food to be out on the glacier for three weeks.

Vatnajökull was as good as unknown, a nameless land back then, like the bulk of the highlands. Only a few people had even gone onto the glacier at the time. A few research missions had been undertaken, but no regular survey or research had been carried out into the accumulation, thickness, and nature of the glacier, at least until the Icelandic Glaciological Society began its regular spring expeditions.

My grandparents stayed in a two-person tent and endured a raging three-day storm that engulfed the tent, meaning they had to stay in and wait it out. Only the very top of the tent glinted above the snow as they were eventually shoveled out. I once asked if they hadn't gotten cold in there. "Cold?" they replied, offended, and laughed. "We were newlyweds!"

I was eleven years old when I asked and for a long time I wondered about their answer. How did getting married make a person warmer?

The highest point of the Kverkfjöll Ridge did not have a name at the time. Dr. Sigurdur Thórarinsson wrote about this anonymous landmark in his article about the trip for *Jökull*, the Journal of the Icelandic Glaciological Society:

> We made for the nameless bunga, a bulge or rounded elevation, which rises from the northeastern part of Kverkfjallahryggur, separated from Kverkfjöll by Gusaskard. Twice on this trip I measured with the barometer the height difference between this bulge and the more easterly Kverkfjall one; according to these measurements the bulge is about 1,760 m high, while Gusaskard is 60 m lower. We collectively named this bulge Brúdarbunga; we expect that name will stick unless another one is found soon.

Brúdarbunga, which means the Bride's Bulge, is 1,781 meters high according to maps: Iceland's fifteenth-highest peak. Its exact coordinates, latitude and longitude, are N 64 35.378, W 16 44.691.

The mountain cabin the group had built the year before was small and cozy, black with a red roof and with sleeping space for twenty. The group was glad to see it had survived the inhospitable winter. The cabin stands to this day, surrounded by black sand, lava mounds, and glacial deposits, near the perimeter of Vatnajökull on the west side. In the guestbook of the cabin at Jökulheimar, you can read this entry:

May 27, 1956
A honeymoon trip, surveying trip, snowhole-digging trip, an airfield construction trip. Twenty-five participants, one fifth women. Arrived here a bit before noon on the aforementioned day, the Festival of the Trinity. We left Reykjavík at 16:25 May 26 with 2 heavy vehicles, "Gusi," driven by Gudmundur Jónasson, and 6 additional automobiles. It was raining heavily all the way east to the Tungna river crossing, we arrived there at 5:30 on the night before the 27th. After two hours we'd brought everything across the river and then we drove up to Jökulheimar in clear weather. The journey was more or less clear, yet we were barely able to make it: the route was practically still closed off by snow when we came in under Ljósufjall, and if we'd been any earlier in the season it still would have been; now, there was little delay.

The expedition's first task was to make a bed for the newlyweds with a soft mattress in the area of the hut which, according to those who examined the lodge very scientifically, had the sturdiest floorboards.

Having eaten hot soup, the whole crew lay down to sleep and most of them didn't wake until the smell of food entered their senses again six hours later. The five women and Bjössi had prepared a delicious supper. Later that evening, a great jamboree started up in honor of the frequently-toasted newlyweds. Cups were drained and sentimental songs sweetly sung. A suitable redness flushed into the bride's cheeks when Úlfar Jakobsson lustily sang a love song. The bride and groom were celebrated with many speeches, and the festivities were a great success. It was heartily wished at the festivities' conclusion, and with voluble assenting, that this should not be the last honeymoon in Jökulheimar. From now on, there will always be a firm bed awaiting newlyweds! You might say this shelter is fit for the purpose . . .

Signed

Jón Eythórsson

I read the diary entry to my grandparents and they grinned like teenagers as I read the part about the mattress on the sturdiest floorboards. My nine- and eleven-year-old daughters were sitting with us and didn't get the joke.

According to the guestbook, the expedition's mission on the glacier that spring of 1956 was as follows:

1. Survey markers to be placed on Thórdarhyrna, east Kverkfjöll, Grendil, Hvannadalshnúkur and Svíahnjúkur both east and west.
2. Snowfall or winter accumulation shall be measured in as many places as possible.
3. If conditions allow, a sledge should be obtained at Esjufjall and the Icelandic Glaciological Society lodge there checked.

4. Travelers, five women and one man, shall be given
   the opportunity to traverse as far across the glacier as
   conditions allow; they are not required to visit all places
   survey markers will be placed.

Gudmundur Jónasson was the tour leader, nicknamed "Mayor
of Grímsvatnahreppur." He drove his snow truck, nicknamed
Gusher, and handled radio and telephone connections with
the outside world. The geologist Sigurdur Thórarinsson took
care of snow measurements; the cook (and bursar) was Árni
Kjartansson and another cook was Mrs. Hulda Gudrún Filippusdóttir. Ólafur Nielsen drove Grendel, a snow truck,
and Haukur Haflidason drove Jökull 1. There was Ingibjörg
Árnadóttir, Grandma Hulda's best friend, and Nick Clinch,
a celebrated mountaineer from America who worked as a
lawyer with the defense force at Keflavík Airport Base.[38] He
went on to be the first person to climb the impassable Hraun-
drangur and later climbed many of the world's toughest
peaks and wrote books about his adventures.

This lively group literally went all across the glacier from
Jökulheimar to Kverkfjöll. At selected places, pits were dug to
survey winter snow accumulations and markers were set up
on the peaks for surveying. At the summit of Hvannadalshnúkur, a two-meter-high barrel was installed and thus the
travelers could stand two meters above the highest peak in
the country. Sometimes they had to travel blind, relying on
compasses and altimeters. Considering the era's technology,
incredible distances were crossed; at the time, the most dangerous crevasse regions had not been mapped properly, or
even found at all. The glacier is more than a thousand meters

deep at its thickest. A whole kilometer of ice. The transceiver reception malfunctioned so they could only communicate outward to the wider world without knowing if anyone could hear. The journey was heavy going and soon they were low on gas out on the middle of the vast glacier and had to act fast to save themselves:

On June 10, four of us made an attempt to get to Grímsvötn to get gas. We weren't able to take the Weasel more than 5 km, due to a lack of fuel, after which we continued on skis. The ride was hellish and the weather worse. We thought ourselves fortunate when we got back to the camp site, having trekked about 35 km, having struggled and remaining unsupplied. We would hardly have found the campsite if we had not been so ingenious as to pile snow cairns at short intervals all along the way.[39]

I show Grandma Hulda and Grandpa Árni a video, a movie Grandpa Árni took on 16 mm film during the honeymoon. One shot was taken down the mountain and another of the group on skis being pulled behind the snow vehicle. It's an enchanting time, adventurous. I ask if they weren't ever worried about crevasses or cracks opening up in the middle of the glacier.

"We had a kitchen tent and made a hole for food scraps and waste. We didn't understand why the hole never filled up. We'd only gone and camped on top of a four-hundred--meter-deep crevasse!" Grandma Hulda laughs. "We were lucky that we didn't lose anyone on these trips. There were no major accidents."

"And you never got lost?"

Grandpa Árni thinks about it. "No, I never got lost. When I was in the mountains, the conditions could get bad enough that I didn't know where I was but there was nothing to do other than dig into a snowdrift and wait out the weather, sometimes for a few days. If no one was expecting me back home, no one would start looking, so I wasn't exactly lost. As soon as it was clear again, I could continue the journey."

"The best thing was hanging off the back of the snow truck and letting it pull us along," Grandma Hulda says. "We could ski down the slopes that way. Once, we hung from the back of the vehicle all the way from Kverkfjöll to Jökulheimar, ninety kilometers. Everyone gave up except for Árni and me."

"But don't you want to go back to the glacier?"

"Yes," says Grandma Hulda. "I would like to go but Árni isn't fit for glaciers these days."

"It's just far too easy today with those monster trucks." Grandpa Árni grins. "It's no fun when things are so easy."

"What's odd," says Grandma Hulda, "is that in the spring I can sense the smell of the glacier and I find myself longing to go back."

"The glacier's smell?" I ask.

"Yes, there's a particular scent to the air in spring, a glacier scent."

"What does it smell of?"

"Well, just glacier! You can only describe the scent by experiencing it. When you're up on Vatnajökull everything disappears; you forget everything. An infinite vastness. An absolute dream."

Sigurdur Thórarinsson's article about the trip in *Jökull* also discussed the research the team conducted:

Most of the measurements carried out by Vatnajökull expeditions in recent years remain unfinished. When these measurements have been completed—glacial thickness measurements, triangulation measurements, and snow accumulation measurements, as well as observations on changes in the Grímsvatn area—we hope that our knowledge of this largest glacier in the country will have been reasonably satisfied. In all likelihood, this glacier will never be fully researched.

When the first trips of the Icelandic Glaciological Society were conducted, people knew that Iceland's glaciers were *living*. A glacier is defined as an ice mass that moves under its own weight. Glaciers are semifluid, and the largest are like a kind of conveyor belt: winter snow gathers on the accumulation areas and becomes ice that flows slowly down into the valleys via outlet glaciers, where it melts. A healthy glacier is in equilibrium. It collects as much snow as it delivers via loss of water. Its income equals its expenditure. In Iceland, a glacier's uniqueness lies in the interplay of fire and ice, with eruptions under the glacier giving rise to *jökulhlaups*: massive, dangerous glacial outburst floods that can achieve the volume of the Amazon in a short period.

A plane landing on a glacier serves nicely as a benchmark for understanding glaciology. The plane Grandpa Árni and his companions dug out was covered in several meters of snow one year after landing on the glacier. Over time, it would have burrowed farther down, then moved slowly along with the flow of the glacier toward the next outlet glacier. There it would have resurfaced as part of an ice tongue, possibly after a thousand years. The ice that calves into the glacial lagoon Jökulsárlón is snow that fell on Vatnajökull's summit at the time of Iceland's settlement over a thousand years ago.

In 2019, Vatnajökull is about 8,000 square kilmeters, roughly 10 percent of the size of Iceland; its total volume is around 3,200 cubic kilometers. If the glacier was evenly distributed throughout Iceland, the land would be covered by a thirty-meter-thick layer of ice. Helgi Björnsson, a glaciologist, told me that the glacier contains water that is equivalent to twenty years' worth of rainfall across all of Iceland; if it melts, the world's oceans will rise by a centimeter. Forecasts of up to a hundred centimeters' rise in sea level mean that the equivalent of a hundred Vatnajökulls will disappear across the world in the next one hundred years. Vatnajökull has already added one millimeter to the surface of the ocean because it reduced by 10 percent during the twentieth century.

Vatnajökull reached its high point, having grown steadily for more than five hundred years, during what's known as the Little Ice Age, a phenomenon regional to the Northern Hemisphere that occurred from the sixteenth to nineteenth centuries. All that accumulation seems to be going in the other direction now, and within just a hundred years, as a result of the worldwide conditions of global warming. Since the turn of the century, the glacier has been reduced by 4 percent, thinning by over half a meter a year. That's almost a hundred cubic kilometers of ice. Other glaciers in the country, Langjökull and Hofsjökull, are retreating even faster. Estimations have Snæfellsjökull disappearing by 2050. Okjökull is already gone. Okjökull was the first known Icelandic glacier to formally lose its status as a glacier. This formerly fifty-square-kilometer ice cap is now one square kilometer of dead ice. Vatnajökull's equilibrium line is at 1,200 meters; at that height, the glacier maintains itself.[40] If a large part of the glacier falls below this mark, snow will cease to accumulate and if that happens, melting will significantly accelerate. That change would be irreversible unless a much colder climate were to emerge. Were a plane to land on the glacier and be buried under winter snow, it would reemerge the very next summer when the snow melts, instead of digging itself farther and farther down. When the glaciers are gone, what is Iceland? Land?

The Icelandic Glaciological Society is a remarkable organization because it's not only for glaciologists, scientists, and others working in glacier-related fields; it also includes a community of amateur scientists and volunteers from the general public. For a poor nation, multiperson three-week expeditions of this kind would have been impossible without a mix of people involved: geologists, mountaineering enthusiasts, adventurous truck drivers, adept builders who could construct shelters on the glacier. It was the enthusiasts from among the general population who enabled the scientists to achieve their goals, and the scientists in turn gave those enthusiasts more depth and purpose. More often than not, work was completed by volunteers (although some of it was funded by the National Land Survey and the Icelandic Hydrological Survey). This was in part the same group of people who founded the Air Ground Rescue Team: mountain and outdoor enthusiasts who dash from work and family commitments to respond to accidents and natural disasters or to search for missing persons. Icelanders have never had the resources to fund a whole rescue team; the strength lies in it being carried out by volunteers.

The Icelandic Glaciological Society's spring trips began in

1953 and Grandma Hulda and Grandpa Árni continued to make these trips well into the 1970s. The expeditions' surveys cover more than sixty years, and they can be combined with weather data and scientists' forecasts of where the Earth is headed.

In 1958, Charles Keeling began regularly measuring carbon dioxide (CO_2) in the atmosphere at the Mauna Loa volcano in Hawaii. Measurements there have shown how the amount of CO_2 in the atmosphere has been increasing rapidly. At the beginning of the Industrial Revolution, the ratio was about 280 parts per million (ppm). By 1958, industrialization had caused a significant increase, to 315 ppm, higher than the highest levels measured on earth for hundreds of thousands of years. Now we have reached 415 ppm, adding about 2–3 ppm a year. The warming of recent years and Vatnajökull's response to it give clear indications of what will happen to the glacier in the coming decades.

Given the rising temperature of the Earth, Vatnajökull's main outlet glaciers will disappear in the next fifty years; the glacier itself will almost entirely disappear within a hundred fifty years. This will happen far sooner if Earth's average temperature rises by more than two degrees Celsius. As my grandparents embarked on their glacier expeditions, glaciers were a symbol of something great and eternal, like oceans, mountains, and clouds. In 1955, many of the country's outlet glaciers had somewhat retreated from their turn-of-the-century position, but Vatnajökull was still something that seemed eternal. An eternal white giant. Its scale of change was many centuries, even a millennium. Vatnajökull is now diminishing on a human scale: 10 percent shrinkage over a hundred years is high speed; 100 percent shrinkage in a hundred fifty years is a disaster. Today, gigantic glacial outlet tongues retreat tens or hundreds of meters annually. For geological phenomena like Vatnajökull to disappear in just over one human lifetime is beyond our comprehension. The size gets lost in the vast buzz.

Spring 2019 marked the first time no one was able to travel up to Vatnajökull from the shelter at Jökulheimar:

> The Icelandic Glaciological Society's spring expedition has taken place annually since 1953. Most often the trip heads out onto Vatnajökull from the west, through Jökulheimar and Tungnaárjökull [. . .] But Tungnaárjökull has retreated so far that the area in front of it has become impassable to vehicles due to the wetness of the ground. No viable route has been found this spring. This is the first time in 66 years of spring expeditions this has happened. These changes are a direct result of global warming. Tungnaárjökull, like other glaciers, is retreating rapidly and what emerges from under it is land that has been covered over by glaciers for at least 500 years.[41]

Watching the movies Grandpa Árni recorded out on the glacier, I once told him he should have shot more images of Grandma. She was only young once, I reckoned, but it would always be possible to film the glacier and the landscape. I was wrong. The glacier turned out to be as ephemeral as a person. The films Grandpa Árni took document a landscape that won't last much longer. If my youngest daughter lives to be as old as her great-grandma, she will still be alive in 2103. It is pure science fiction to think so far ahead. By then, Skeidarárjökull will be long gone, Langjökull mostly gone, Hofsjökull, too.

Where the glacier once touched the sky, there will be only sky. Our grandchildren will look at old maps and try to imagine a mountain made of frozen water. They will try to understand its nature. A thousand meters of ice filling a whole valley? They will draw lines in their minds and strings between peaks, imagining a glacier that is so thick the world's tallest towers pale in comparison. They will point to the air and say: "That's where the campsite was. Up there, right under the clouds, that's where the two of them pitched up on their honeymoon." Our grandchildren will try to conjure mental pictures, try to envision a snowmobile traveling across the sky towing singing skiers. It is impossible to imagine what else will be happening in the world at that time. And impossible to say what words and concepts will be used about our time. It depends on us, on what we do right now.

In the summer of 2019, I was given the strange task of writing a memorial for the Okjökull glacier, the first of hundreds of glaciers Iceland will lose over the next two hundred years. It took me quite some time. I wondered who I was addressing with the words on the plaque; I wondered at the absurdity of the task. How do you say goodbye to a glacier? In the end I came up with this:

Ok is the first Icelandic glacier to lose its status as a glacier. In the next 200 years, all our glaciers are expected to follow the same path. This monument acknowledges that we know what is happening and what needs to be done. Only you know if we did it.

The mother of the universe,
white as rime

The Himalayan region is home to 46,000 glaciers. These are to be found amid the world's highest mountains, many protruding as high as eight thousand meters. They are scattered throughout the mountain range, hiding in valleys and cirques under precipitous mountain slopes. In total, the Himalayan glaciers are only about four times larger in area than the Icelandic glaciers. They cover a total area of about forty thousand square kilometers, but the volume is very similar to the volume of all Icelandic glaciers, about four thousand cubic kilometers.[42]

Iceland's glaciers in many cases reach down to sea level, and one might expect such glaciers to be more sensitive to climate change than glaciers far up in the Himalayas. Recent research, however, shows that high-altitude glaciers, where the state of winter might seem eternal, have retreated to a similar extent as the Icelandic glaciers. The surface area of these glaciers tells only half the story, since some glaciers are diminishing in thickness by a meter per year. The receding of glacial outlet tongues in the Himalayas has led to unstable glacier lagoons forming behind weak glacial moraines. These moraines crack, causing tremendous floods with related casualties in towns and villages that lie close to the rivers.

As the Icelandic glaciers retreat, the flow of glacial rivers increases. This is good news in the short term for energy companies that harness hydropower, until the flow decreases again and becomes tantamount to precipitation. The glaciers affect weather systems, drawing in precipitation so that it mostly falls on the south side of the larger glaciers. Precipitation will likely fall in more northerly spots if glaciers disappear. In the Himalayas, the flow of glacial water is expected to increase over the next few decades. The problem is that while most of the Icelandic rivers flow fast and free to the sea, the rivers in the Himalayas provide the livelihoods of billions of humans and the foundation for a diverse biosphere from mountain settlements to estuaries. As water increases due to melting, a false prosperity can emerge, as if you were to take all your money out of the bank or burn through your supply of winter hay in the first month. There's something vicious about the problem of a temporary increase in water flow improving the quality of life, the soil vegetation, the groundwater status, and even electricity production—all before the rug is pulled out from under the feet of millions of people. When the glaciers have gone, another system, a more finicky one, will come into being, marked by excess water during the monsoon season and water shortages during the drought period—these droughts might well be longer and more extreme due to the rising temperatures. The services provided by glaciers cost nothing. People are beginning to understand why the glaciers in the Himalayan region have been considered sacred.

Lonnie Thompson is a professor of glaciology in Ohio. He probably has the most knowledge of anyone on earth of the world's high-altitude glaciers, their history, and future evolution. Information he has collected over the last few decades has provided the world with data on which climate science bases its predictions. He has taken ice cores from glaciers around the world and directed more than sixty expeditions in sixteen different countries. By "expedition," he does not mean quick glacier trips, but monthlong stays on the world's most remote ice sheets. He has taken cores from glaciers up to seven thousand meters high. In the tropics, this can be a daunting task, transporting ice from the highest altitudes, through jungles, to a laboratory in the United States.

Glaciers are frozen manuscripts that tell stories just like tree circles and sedimentary deposits; from them, you can gather information and create a picture of the past. Glaciers store histories of volcanic activity. They store pollen, rainwater, and air that reveal the chemical makeup of the atmosphere tens of thousands of years back in time. They are important sources of details about vegetation and precipitation of the past.

Looking at the past millennia, changes in the world's glaciers have not always been unidirectional. The southern hemisphere and the northern hemisphere have not been in rhythm; glaciers have grown in certain areas and decreased elsewhere in accordance with local climates and local fluctuations. But as of today, almost all glaciers everywhere in the world are simultaneously melting.

"I never would have believed that a glacier could disappear so fast," says Lonnie. "Not in ten years. If anyone had predicted such rapid melting, I wouldn't have believed them."

On the Tibetan plateau, 680 glaciers were studied: 95 percent of them had retreated. In Alaska, 98 percent of the glaciers are rapidly diminishing. Climate change deniers exhibited thirteen glaciers that had, by contrast, grown as a result of increased local precipitation, claiming them as proof that nothing was happening to the world: "news" of "accumulating glaciers" spread around the globe. These exceptions were paraded; claims were made that the earth wasn't warming but, on the contrary, cooling. Media personalities appeared reconciled to presenting non-scientific information as "news."

"The world's glaciers are retreating everywhere but humanity is closing its eyes to the consequences," Lonnie continues. "Millions of people live in places that will become virtually uninhabitable within decades."

Lonnie tells me that, looking ahead, the life of almost all the glaciers outside the Arctic will end in this century. Like most of the world's leading climate scientists, he had been living through his own dog days, depicted as a conspiracy theorist, a communist, an extremist. Politicians, especially in the United States, did not seem to understand the words or data he used, nor to sense its importance, even though it touched upon the foundations of life in general and the specific lives of millions. He watched politicians surround themselves with people connected to oil interests. Self-appointed experts would make statements that had nothing to do with science: "An increased amount of CO_2 in the atmosphere is good for vegetation." There were politicians who listened to and even seemed to understand the information but still did nothing.

"And what's going to happen?"

Lonnie replies, "The most recent papers looking at sea level rise are looking at a likelihood of a one-meter rise within less than a hundred years, and some of them up to two meters. If you look at a two-meter rise, you are displacing hundreds of millions of people because unfortunately many of our big cities are built right at sea level. We're faced with determining where all these humans are going to go. If you go back two hundred years there were parts of our planet that we didn't occupy. You could migrate, you could go somewhere, but now human beings occupy almost a whole planet. We are seven billion people on the planet and the question is, where do they go? And who is responsible? People are already being impacted in many parts of the world. In Peru, two-thirds of their twenty-eight million people in the country live in the coastal desert, dependent on rivers that originate in the Andes at the base of many of these glaciers that are getting smaller every year. On the Tibetan Plateau and in India, you can see caves in the mountains where people used to live two thousand years ago. Today, 80 percent of people's water there depends on melting glaciers in the dry season and in the next hundred years they may have to move out of this region because there's no water. You look at some of these countries and there's always the risk of a 'failed state,' if suddenly you have a disruption in the water supply. Especially since they cross borders and if you are in one country that tries to build a dam you're going to deny water for people downstream. From a geopolitical point of view, these are rivers that supply water to three nuclear powers."

"And how do you see the future?"

"At some stage in the future we'll pass a threshold where you'll start seeing dramatic drops in water discharge during the dry season. How do you adapt to that and do that in a peaceful way? If you live in areas that depend on water coming from glaciers, from what are essentially nature's water towers storing the snow in the wet season at no cost and releasing it gradually in the dry season, at no cost—that's going to change. How are you going to adapt to these changes and still have water supplies and electricity and the things that make life possible?"

His scientific perspective is troubling because of how gloomy it is.

"Is there any hope?"

"For our research projects, our field teams are international, Nepalese, Chinese, Russians, South Americans, and Peruvians from the Andes. We're able to go and recover ice cores down to bedrock and live on these mountains for six weeks at a time and yet this international group can work together to accomplish the objective we're there for. So I believe when the day comes when humans realize that the only way that we can solve this problem is by working together, then we can actually come together and do that. But I also believe that we won't until our backs are to the wall, until it's such a crisis that we have no other choice. And then we're pretty good, I think; I think humans can work together. There are all kinds of examples of this in our history, you can look at World War II: when a danger is recognized and perceived to impact billions of people, we can focus and dedicate resources and change direction. And it really does mean changing our lifestyles. I just hope we don't have too many people suffering before we actually become serious about change."

Farewell to the white giants

In the future, glaciers will be an alien phenomenon, rare as a Bengal tiger. Having lived in the time of the white giants will become swaddled in a fairy-tale glow, like having stroked a dragon or handled the eggs of the great auk. Glaciers will certainly be found in the Arctic, Greenland, and Antarctica for a few thousand more years, but probably not in the Alps and the Andes; they will disappear in most parts of the Himalayas and Iceland. People will ask, how were glaciers described at the beginning of the twenty-first century?

I was not as familiar with glaciers as my grandparents were. I had seen them from afar and gone up to Snæfellsjökull in winter, but winter glaciers are nothing like summer glaciers and outlet glaciers are quite different from an ice sheet or a minor glacier.

And so we planned to cross Skeidarárjökull, where it heads south from Vatnajökull, one of its major valley glaciers. It was the end of July 2012; all the winter snow had melted and all the cracks and shapes in the ice were as clear as they would get. This was actually our second assault on the glacier. A few years earlier we had headed up there in pouring rain and pitched our tents on a low gravel bed. When we woke up in the morning, pools and springs had formed under the campsite. It was almost as if someone had struck the ground with a magic wand, causing water to well up out of little bulging eyes; people woke drenched in deep puddles and so we turned back home.

Now the plan was to camp at the edge of the glacier, fairly high up, and cross the glacier in one long day trip, a total of about twenty-five kilometers. Then we would camp on a green terrace, one of the most beautiful campsites in the country, and make another long day trip into Skaftafell National Park.

We woke up in crappy weather; the tent was shaking in the wind. It was warm down inside our sleeping bags but shiveringly cold once we crawled out of them. We packed up quickly and set off in spite of the poor visibility. At the edge of the glacier we ran into some hikers who had crossed the glacier during the night. These were a French father and son and the father's friend; they were cold and dazed, almost in shock after the night's hardships. They had lost their way and come across a crevasse that could not be traversed; it led them astray so that they ended up too low down, where they got into a maze of deadly deep crevasses. And so they went back and forth blindly in the fog and rain and darkness. Ten hours of walking became twenty hours. They feared for their lives and pitched their tent as soon as they stepped off the ice, bone-tired and relieved they had reached safety.

We ourselves tramped though slushy mud at the glacier's edge, where glacier meets land. We tried to avoid the quick-sand that forms when melting glacier ice seeps water into the sediment. The glacier was black with sand for the first part of our journey and on the ice we could make out strange objects, flat stones on thin ice pillars, like works of art made by aliens. The weather was slowly clearing, until we started to see before us an endless breadth of tussocks, as though innumerable white turtle shells stretched out as far as the eye could see.

We saw that in the middle of the mountain slope on each side of the glacier was a light stripe that marked its surface level as it had been just a few years ago. In many places high up in the cirques, we could discern so-called dead ice, floes still hanging there in the rock after the surface had subsided. It tests every sense of one's brain to imagine a glacier's surface thirty meters above one's height, the height of a ten-story building, extending it in the mind, as though it's a vaulted ceiling over the entire expanse, reaching one edge of the glacier to the other and a whole kilometer out onto the sand.

Our path continued and now it was as if each of the tussocks were individual scales, with the outlet glacier the tail of a white dragon. We came upon something that resembled a black sand pyramid and then more pyramids gradually appeared until we entered an entire forest of black pyramids in the middle of the glacier. The evaporation from the sand cones emitted fog veils; little streams trickled between them, creating a micro-landscape, almost like a bonsai landscape, little mountains and little rivers and little towns, and we were mesmerized by the shapes and the beauty. Between the pyramids, streams ran like little waterslides. It was tempting to take a ride but if we followed them, we'd end up in a glacial hole. These gullets were white holes that became blue holes and black holes that extended as much as three hundred meters down to the bottom; it was vital to take special care around them. The thought of losing one's footing and disappearing down into a hole was the stuff of nightmares. The only parallel my mind could conjure was the sandpit in *Star Wars* where the gigantic, wormlike Sarlacc lived.

In the book *Solaris* by Stanislaw Lem, astronauts float behind a mysterious planet and try to figure out its nature. They hypothesize that the planet has some kind of self-awareness beyond what the human mind can comprehend. On the surface of the planet is a kind of ocean of yellow foam that takes on familiar shapes the astronauts try to interpret and understand. They wonder if the planet is sending them messages.

I tried to interpret the glacier's shapes, the pyramid forest opening out into a streak on the surface that resembled a two-lane highway. In the middle of the "road" was a black line as if to mark the lanes. The surface was smooth and level; one could have driven at seventy miles an hour as far as the eye could see. I involuntarily looked both ways as I walked across the "road" and wondered if the glacier was giving me a signal. Maybe it was telling me that somewhere between the pyramids and the motorways something had gone wrong.

I lay down on the cold ice and put my ear to a narrow crevasse that seemed a whole eternity deep though only a few inches wide. The ice in the wound was as clear as crystal. I looked at the veins and bubbles in the body of the glacier, which created a strange three-dimensional feeling. I heard how the water gushed far down in the quivering space like a dark bass, water dancing somewhere deep down at the bottom, like a giant xylophone, a rock harp, an ice harp. The glacier's swan song.

Now that the glacier is changing faster than ever before, I feel within myself a paradox. My being on the glacier comes from advances and technologies, the production and mass extraction of Earth's resources. By the time humans were able to cross glaciers, to count the nesting places of crocodiles, and to study the song of the humpback whales, we'd grown so strong and expanded so far that what we were finally able to measure and understand had already started disappearing.

In documentaries, melting glaciers are a dramatic spectacle: gigantic ice chunks crash and rumble as they calve into the sea. But a dying glacier is actually no more dramatic than the normal changes of the spring season. Ice melts in the heat and the sun, forming streams that frolic and splash. In fact, a dying glacier is more a sad, frail sight, disappearing quietly. You could call the situation "silent spring," had Rachel Carson not already used that phrase to title her book on how insecticides affect nature. And after spring will come summer. The long global summer.

The place names on Vatnajökull store memories of the changing environment. Breidamerkursandur, meaning "wide forest sand," recalls a forest there before it became a black sand desert. Under the sand, we can find the thick stumps of three-thousand-year-old birch trees from a time when the local Nordic climate was warmer or as warm as it has become today. The wide forest turned into a wide desert after the advance of glaciers in the Little Ice Age. The self-sowing birch is beginning to breed itself once more. Breidamerkursandur will probably become a wide forest once again. On Skeidarársandur's endless stretch of black sand, the largest self-contained birch forest in Iceland is beginning to form.

Could you really call the largest forest in Iceland *skeidarár-sandur*, "boat river sand"? The forest would be named after a vanished glacial river, after black sand deep under the forest floor.

Ice, gravel, and sand emerge from under the glacier, a new land that has been frozen for hundreds of years; at the glacier's edge you have to tread carefully on the ground. It's as if the land is neither ice, nor water, nor sand but all three at once. The transformation has an intermediate step: chaos, as in the prophetic poem "Völuspá," about the beginning of creation:

> The sun knew not
> where her hall stood,
> the moon knew not
> how mighty it was
> the stars knew not
> their stations.

Chaos is not confined to the glacier's edge. No one knows how the land will lie when our way of life has caused the world's glaciers to become water, our coastlines to become sea, and our arable fields to become deserts.

When I turn ninety, I will show my thirty-year-old grand-child pictures of Skeidarárjökull on a projector screen. They'll see a glacier that three generations of my family had the opportunity to get to know before it vanished. When I take a photo of a glacier, it's like I'm recording and preserving an old woman singing an ancient lullaby. After a thousand years, people will peer at the pictures like rare, ancient man-uscripts and try to understand what we were thinking.

The god in the steam engine

Take the hair from my tail and lay it on the ground.
It will make such a vast bonfire that no one can pass it
except the flying bird.
—Búkolla, the magic cow (Icelandic folktale)

One might suspect I'm trying to lure people into a cult: the Guardians of the Sacred Cow. In a time when we worship machines, technology, fashion, and brands, there's pressing reason to form groups that directly worship Mother Earth, the birds, the forests, and the ocean. We could adopt a little prayer:

> Our Mother
> You are the earth
> Your Kingdom is here
> Holy is your water
> You grow for me my daily bread
> Until I merge into you
> The eternal cycle of life and splendor . . .

And yet, sadly, I am not the fruit of the earth and we are no longer 70 percent water. We're 90 percent oil. Earth's inhabitants numbered seven hundred million when we started exploiting the coal and oil that gave us superstrength. We persuade ourselves that the basis of our society is idealistic but when push comes to shove it is fuel that undergirds everything. We're now seven billion strong, having cheated the system and dug our way down to ancient sedimentary deposits, strata from long-dead organisms. Like exorcists, we disturbed their infinite sleep, pumping them back to the surface, kindling fires and harnessing hundred-million-year-old sunshine as it lay dormant in the Earth's belly. We bent fire to our will. We drove ships right into storms to shovel up fish, we threshed grain, we built cities, and we flew above the sea in metal dragons, all thanks to the power of combustion. All this has increased slowly but surely for more than two centuries, and never more so than at this very moment.

Although the atomic bomb is the chosen benchmark for marking a new geological era, that geologic turning point, that new geological century, might be dated earlier: the geological era of the Anthropocene, which will be known by Earth's warming, by glaciers melting, by rising sea levels, by acidifying oceans, by species going extinct, can be traced directly to James Watt, born in Greenock, Scotland, in the year 1736.

Watt was a skilled mechanical engineer who designed measuring devices for the University of Glasgow's laboratories. He revolutionized the steam engine in 1765, formally taking over Prometheus's torch. Watt hid the fire inside a metal cylinder in which water was heated to boiling point. He managed to harness the resulting steam energy to turn a piston that propelled machines that pumped water, shoveled in coal and iron ore, and allowed humans to create even bigger and more powerful machines, which in turn created even bigger machines. With them, man could dig deeper into the Earth and stretch railroads across entire continents so that screaming locomotives could inspire poets who praised the new dawn of humans.

Oil was first extracted around 1860; the internal combustion engine followed shortly after. We learned to fly and we headed out into space and humanity has grown mightier and mightier over the past 250 years.

Watt allowed us to tame fire; gradually we learned to hide the flames better, isolating the heat, smoke, and soot until fire has become invisible under the glossy hoods of air-conditioned automobiles that go cruising around the world with their upholstered seats and pumping music.

The fire Watt ignited has become a massive, unstoppable inferno, bigger than any forest fire that has ever blazed on Earth. In total, we have burned hundreds of billions of tons of coal, and as far as we knew, these mountains of coal had simply evaporated, vanished. But they haven't. The coal and oil became carbon dioxide, CO_2. Scientists today can measure the amount in the atmosphere and compare it to cores from Antarctica and from Greenland's glaciers. That reveals what the invisible mountain of CO_2 looks like compared to the levels during the last 800,000 years.

When Watt ignited his steam engine, the proportion of carbon dioxide in the atmosphere was 280 million ppm; the levels have since reached 415 ppm, the highest in three million years.[43] What about volcanic eruptions? some ask. Aren't we humans pipsqueaks compared to Earth's volcanic activity? Unfortunately, that's not so. Earth's volcanoes are estimated to release on average about two hundred million tons of CO_2 a year, while humanity releases thirty-five *billion* tons a year. The fire we burn is almost two hundred times greater than all the combined volcanic activity on Earth.[44] Yet we go through our day without actually seeing fire or smoke. We see and perceive volcanoes, their ferociousness and their thundering din, but we don't see that we are Earth's largest volcano.

When the Eyjafjallajökull eruption began in 2010, putting a stop to European air traffic for six days, the volcano's emissions constituted about 40 percent of daily European air traffic emissions—about 150,000 tons per day.[45] By halting all air traffic, it prevented 300,000 tons per day of emissions, and so became the first environmentally responsible eruption in history. Beneath all we do there's a fire blazing; traffic advances like glowing lava. If we divide the 100 million tons that humans emit every single day by the 150,000 tons that our volcano emitted, we get the devilish number 666. The emissions of the earth's inhabitants are like 666 Eyjafjallajökull eruptions, day and night, all year-round. If we convert U.S. emissions into volcanic eruptions, one hundred volcanos like Iceland's Eyjafjallajökull erupt there every day, every night—practically two volcanoes per state.[46]

We are the eruption, but we don't see flames when we look in the mirror: everything is so well designed, so invisible. If the cars on our highways displayed their fires on the outside, the conflagration we kindle in order to get to work would be evident. We would see a distinct difference between an electric car and a gas-powered car, but the electric car also conceals flames and emissions: the fire that goes into production, the flames of the ship transporting the car between countries. If fire were to rise from the bodies of cars, we could see during the shadows of midwinter the mighty lava flow this frenzy of traffic constitutes. We can see forest fires and burning high-rises. On the news we see tankers burst into flame or an accident causing oil reserves to catch fire, without thinking through how all this oil was meant to burn anyway, just not all in one place. We do not perceive our everyday disasters.

We do not see fire; we rarely see coal or oil. We're frequent flyers but we have no idea about the size of the bonfire that could be ignited with twenty tons of jet fuel. We buy our airline tickets online but we never have to check-in the oil barrels that will carry us out into the world. Take the time I went to a two-day poetry festival in Lithuania, a journey of around 1750 miles, the same distance as Chicago to Los Angeles. A barrel of oil holds about forty-two gallons, so a single airline passenger burns through about three-quarters of a barrel on such a flight: up to one gallon every sixty miles. I'd have needed one-and-a-half barrels to get me there and back.

The oil becomes CO_2, an invisible, odorless gas. But when it comes to the natural world, nothing gets used up; energy can be transformed from one form to another, but it cannot be created or destroyed. The oil burns but doesn't vanish, doesn't diminish; quite the contrary. One ton of oil turns into 2.3 tons of CO_2. Forty-two gallons of oil thus amounts to 350 kilograms of CO_2. Approximately 50 tons of CO_2 for one hundred and fifty passengers. "Hello, how can I assist you?" "We can't be bothered to fly to London this weekend but could you just sell me two barrels of oil instead? We're going to have a high old time lighting a big bonfire in our backyard."

I pump gas into my car; I could just as well fill it with data, the liter counts rolling down the screen along with the price.

But the price isn't accurate: I pay the cost of pumping the oil out of the ground and I pay the company's profits but I do not pay for the consequences of putting all this CO_2 into the atmosphere or for the 400,000 square kilometers of land that could sink underwater this century. Someone has to pay to remove all this CO_2, either by planting billions of trees or by inventing new technologies on a massive, totally unprecedented scale to recapture the CO_2 from the atmosphere. Not paying is not an option. If we don't, future generations will have to pay with Earth's biosphere and with their lives along the way.

This fire is invisible and it becomes "nothing"; the smoke simply evaporates. It would be instructive if everyone had to store the oil barrels they use, if we saw the world that way. Our family's trips abroad over the last ten years amount to a hundred barrels of oil. I imagine stacking them outside my home. I imagine the stacks when everyone has a hundred barrels in their backyard. I imagine a community where a pile is a status symbol, an unmistakable sign of owning a powerful car and having made the most trips to Bali. Someone who has driven the family car 100,000 miles has burned over 2,500 gallons of gas, about sixty-five barrels of oil. A five-person family with two cars guzzles around 200 barrels in a decade.

The barrels get added to the space cars take up: an average of 50 percent of city land is roads and parking. The barrels could be painted like clouds to remind us that although oil burns, it does not go away. The barrels should be black like thunderclouds.

The stack is massive but how massive is the blaze? In October 2018, the world's oil production exceeded one hundred million barrels a day for the first time—the exact same week the United Nations Special Report on Global Warming of 1.5 degrees Celsius gave a stark warning to mankind about the disastrous consequences of a two-degree rise in global temperature and the importance of not exceeding 1.5 degrees Celsius.[47] If the oil in a hundred million barrels were turned into a river, the flow would be about 185 cubic meters per second. This is as much as the average flow of Dettifoss in northern Iceland, Europe's most powerful waterfall, visible in the opening scenes of Ridley Scott's *Prometheus*. Watt stuck his Promethean torch inside a steam engine and the result has emerged two hundred and fifty years later. An everlasting, ever-flowing carbon-black waterfall that plunges over the cliff face day and night, all year long. Close your eyes and see it for yourself. Light it on fire and look at the flames rising to the heavens.

The world's emissions are not only from burning oil, but also from gas and coal. Annually, about four billion tons of coal burns. About six hundred kilograms for each person on earth. Twelve billion tons of carbon energy burned annually that is sourced from oil, gas, and coal, turning into more than thirty-six billion tons of CO_2. Thirty-six gigatons.[48]

Imagine all that as a single engine powered by a whole river of oil. Imagine the power of the engine, the injection pump, the cylinders, and the exhaust pipe. This is our life-machine. This is industry. The world's combined gasoline engines are lethal. That's a fact, unfortunately. The party's gotten out of control and must come to an end as soon as possible. According to the latest United Nations reports, we need to extinguish all these fires by 2050, otherwise things

will go very badly for us. We are searching frantically for whatever might lead us toward a life in balance with the Earth, or else it'll shake us off. We must connect ourselves to sources of energy that work in concert with the sun, the waves, and the wind.

To say that we live in a mythological time is not an exaggeration. World leaders meet and *talk about the weather*. It is near-revolutionary that they meet to discuss how they are changing the weather. They talk about hurricanes, thunder, and lightning, the oceans' surface, about deserts and the future of glaciers. They discuss whether to let the Earth's temperature rise by 1.5 degrees Celsius, two degrees Celsius, to let it rise uncontrollably. It is literally a tragedy: our weather gods are as vain and human as the gods of Greek tragedies. People who side with Team Black Gold whisper in our leaders' ears and distract them. It is tragic because Cassandra has figured out where we're headed and what to do about it. The prophecy's clear yet the oil spokesmen insist: "But what about industry? What about profits, growth, the market? What about the labor force, big business, employment figures, election funds, investment and return? If you turn against oil, we'll choose leaders who worship fire." But when people lose all they have to floods, droughts, and forest fires, they'll turn against the people who created the problems, those very same people.

Let's not, however, paint too bleak a picture. We must remember to be thankful for the superpowers oil provides. The energy in one oil barrel equals as much as ten years of a healthy person's labor. How long would I spend hauling a two-ton car containing my whole family along the five-hour drive to Akureyri, Iceland's second city? How long would I spend mowing, shoveling, or plowing the fields? That's no problem with fifty liters of oil. Oil is for us what the Devil was for the medieval Icelandic priest Sæmundur the Wise. Sæmundur was famous for doing deals with the Devil, tricking him into washing his laundry, harvesting the fields, even getting the Devil to carry him home to Iceland from France in the shape of a seal so Sæmundur didn't even get his coat wet. Oil eases our chores, relieves our burdens. Right now, it's doing hundreds of people's work; like the Devil for Sæmundur, it moves us across the ocean. But at the end of every tale, Sæmundur finds himself in deep trouble: the Devil demands payment in the form of Sæmundur's soul or his unborn child.

Oil has lifted humankind out of backbreaking labor, away from shortage; the Western countries first, billions following in their wake over the last few decades. The golden age of humanity. Oil means work and school, means better health, longevity, food safety, late-night TV, summer vacations. The quality of life for the average working-class person is better now than it was for seventeenth-century royalty. And Louis XIV couldn't fly to Tenerife; his teenage son didn't have a smartphone with GPS, a camera, and movie-making capability in his pocket. He never tasted a kiwi or Skyped someone in China. If we get a disease, chances are there's a cure. Energy has swelled our numbers even as the prevalence of famine has decreased; despite wars over oil, war has decreased around the world.[49] Oil has allowed us to create beauty. Suburban teenagers can come together with their instruments and play Beethoven's Fifth Symphony. A billion people now experience the same thing as Papa Árni and his siblings did when they moved into the workers' apartments on Ásvallagata. Deep gratitude for their first stove, luxuries such as running toilets and hot showers, even a spare room to develop photos and perhaps extra energy for leisure, exercise, travel. The problem is not that people have risen out of poverty, but how we have toppled into overconsumption and waste. How we live in a system where most of what we create, think, and make ends up being burned as fuel or becomes landfill. The Earth cannot withstand all this fire; there's no way it can withstand all this consumption. It's choking in garbage. Oil is for us what the Devil was for Sæmundur the Wise. And now the Devil's demanding his pound of flesh. That is where we now find ourselves. Suddenly we see the graphs, the outlook for 2100. We have made life easy in the present by sacrificing unborn children in the future.

Today, the global burning of fuel looks something like this: China is by far the largest polluter, with around 28 percent of global carbon dioxide emissions. The U.S. releases about 15 percent, India about 6 percent, Russia about 5 percent. Germany around 2 percent and the larger Europe nations and Brazil each account for about 1 percent of the world's total emissions.

If we look at which sectors these emissions come from, 25 percent comes from coal power plants, central heating, and electricity generation. Then 24 percent is due to food production, logging, and land use; 21 percent is from industry. Transport as a whole is huge, accounting for about 14 percent of total emissions; there are around a billion private cars in the world. One of the leading causes of emissions from industry is cement production, which accounts for about 6 percent of the world's total emissions. If we examine transport emissions more closely, about 2.5 percent is from international aviation. Everywhere we set foot, we find hidden fires and emissions: from our cars, from our laptops, at work, at dinnertime. That sirloin is on fire.

China is the world's biggest polluter in total but the United States is bigger if you consider carbon dioxide emissions per capita, amounting to 16.5 tons a head. Emissions per capita in Iceland are estimated at 14 tons annually.[50] That amounts to one of the highest in the world despite our "clean" energy, geothermal water supply, and hydropower, providing five times more than our daily need. The average Indian emits only about 10 percent of what an Icelander emits, while the Chinese have overtaken many Western nations in CO_2 emissions. They are approaching the Germans, releasing more per capita than the Spaniards, Swedes, French, and British.

Oil consumption is a fundamental part of the world's inequality. World oil production was approximately thirty-three billion barrels in 2017, amounting to more than four barrels per person. However, these barrels are extremely unevenly shared. About one-eighth of humankind does not have access to oil or electricity, while many nations burn oil primarily to create products that get exported to richer countries.

According to Oxfam's 2015 report, the richest 10 percent of the world's population is responsible for around 50 percent of CO_2 emissions.[51] So while population growth is a problem, the more significant problem is overconsumption and prosperous nations' irresponsibility. Anyone who belongs to the richest 1 percent of the world's population is likely to be responsible for emissions equivalent to that of 175 individuals from the poorest 10 percent of the world's population. But the consequences of climate change come down hardest on the poorest, who cannot afford to protect themselves and who have less freedom of movement. Who will compensate Bangladeshi residents for damage they themselves did not cause when their lands flood?

In order to achieve the goals of the Paris Treaty—that global warming should not exceed 1.5 degrees Celsius—CO_2 emissions must be reduced to zero by 2050. In order to succeed, we will also need to invent technologies that remove CO_2 from the atmosphere in quantities that are equal to all today's emissions. This is one of the biggest challenges humankind has ever faced. What is being proposed is an unprecedented turnaround in the world's energy mechanisms. And 2050 is exactly as far into the future as 1990 is in the past. Since 1990, emissions have increased from twenty-two gigatons to thirty-six gigatons. That's a 60 percent increase.

To get emissions down to zero in thirty years sounds like an unmanageable task. Like constructing a time machine, thwarting gravity, or inventing a pill for bringing someone back to life. No one knows whether it's technically possible to capture thirty gigatons per year. The technology is at an early stage and no one has figured out buildings or infrastructure that could enable us to achieve our goals.

Even if we reduce emissions by 50 percent, our problems will still have increased if we do nothing to remove the carbon dioxide already in the air. If we don't succeed in that project, the Earth will continue to warm, the glaciers will continue to melt, and the sea levels will continue to rise, submerging cities and coastal areas.

The market value of a hundred million barrels of oil is about six billion dollars, assuming a $60/barrel price for oil. We therefore burn approximately $600 billion a day. If anyone thinks changing our sources of energy will be a simple matter, one that will go unopposed, they have another thing coming. Oil underpins entire economies; the effects of energy shifts will be immense for those relying on this sector for profit and gain. That six billion dollars a day won't be surrendered casually, not without a struggle. Millions of jobs and tremendous resources are at stake for rich and wealthy individuals who have direct access to world leaders.

This problem, and its solutions, cannot be underestimated. If we succeed in solving it, if oil becomes worthless or even banned, we'll have pulled the rug from under the feet of entire economies. Russia could crumble, Alberta in Canada would go bankrupt, Saudi Arabia and Qatar could become the next Syria, and Norway would enter a major depression. If that happens, the problem won't be theirs alone, given that oil production has chiefly served Western interests. Yet the only way forward is to recognize the problem and work to free ourselves from this vicious cycle. The world's population will increase by two billion by 2050, according to estimates; we must expect a linked increase in emissions simply from the need to feed and clothe these humans. We're in a tight spot, no doubt about it. We'll perish if we don't put out the fires. We might perish if we do.

But it's not like only the Earth and the sky are breaking apart. There's more. The Earth and the sky are dying and so, too, is the sea. Climate change discussions did not address oceanic change for a long time; for quite a while, people simply hoped the sea would take in as much CO_2 as possible. But now the astonishing rate of change in the ocean's acidification has come to our attention.[52]

To prevent global warming of 1.5 degrees Celsius, the calculations suggest humans have a maximum aggregate "budget" of three hundred to eight hundred gigatons of CO_2 emissions.[53] Assuming thirty-six gigatons as the annual emission rate, our quota will be exhausted after less than twenty years. After that, we mustn't release anything. But as things stand today, we're headed for a three-to-four-degree rise in Earth's temperature by the century's end. Such warming is on the same scale as a nuclear winter. Rising temperatures will increase the forces of hurricanes and storms, leading to more extreme climate scenarios, including droughts and floods that damage crops and wash away arable land. Refugee migrations will likely increase from uninhabitable parts of Africa, China, India, and the Middle East; forests in Australia, America, the Amazon, and the Nordic countries will erupt into flame; the Siberian permafrost will thaw and release methane gas, and this will increase the pace of global warming still further. By the end of the century, the situation we see now around the Mediterranean will be minor compared to what happens when hundreds of millions of climate refugees are on the move. That is: war, death, and destruction. The aim is not to stoke fear. But the paradox is that the problems won't get solved unless we first agree they are problems. We need to be burning with fear yet simultaneously believe the situation can be fixed.

Of course, one can bind hope to denial: everything will be all right. No realistic research suggests so. Those who deny human-caused climate change have become increasingly akin to flat-earthers. Earth's climate has of course varied; yes, giant volcanic eruptions have had an impact on geologic history. But as things stand now, humanity has erupted and eruptions of this magnitude have always led to disaster.

It's easy to slip into hopelessness, irony, apathy. Compare imagining a solution to global warming to having to envision flying a jumbo jet in 1905, a cure for AIDS in 1985, a trip to the moon in 1940. Extinguishing the fires is our major task. Turning away from the black sun and its fossilized black deposits, reconnecting with the Earth and the glowing sun above our heads: that's this generation's role and the role of our children. The stakes? Life on Earth—their lives.

I have no other alternative than to believe a solution is possible. But for it to be so, people must crave a solution as passionately as they craved to fly, to cure AIDS, and to get to the moon. Scientists point out that in order to create a new global energy system, we'd need to invest about 2 to 2.5 percent of the world's GDP over the next few decades. This includes making our transport fleets electric, conserving energy, shifting domestic heating energy sources, building wind turbines, harnessing solar energy, and utilizing thermal energy.[54] The percentage seems surprisingly low given how large and important the project is. By comparison, the British devoted 50 percent of their national production to fighting World War II. If the oceans and the world are at stake, 2.5 percent of GDP is nothing. To send four people to the moon, the United States donated 2.5 percent of national product over a ten-year period.[55] The world's average defense contribution amounts to about 2.5 percent of global national production. Precedents exist. But no army can defend its people against our greatest threat: man-made climate disasters.

Casting James Watt in such a negative light is ungrateful, never mind linking him to the defeat of humankind. He's the father of us all: we owe Watt our lives, for the way he hacked Mother Earth, found a loophole in her system, let us suck the oily black milk from deep below our feet. We're flying high on a million-year-old photosynthesis, tearing ourselves away from seasons to live forever in summer. We can get fresh strawberries any day, all year-round, and buy products specially designed to expire quickly, efficiently creating a need for the same product over and over. The earth is not easily provoked; it's been a long time coming, and now the earth has stirred. It's trying to shake us off. Oil is our life; it is also our death.

When I was younger, the global arms race was at its ecstatic climax and my generation was feverish with terror at the idea of nuclear war. I thought I'd never make it past fifteen. Was my anxiety unnecessary or did a whole generation's worries ensure that the world survived? If we look at the effect burning oil has had, can we apply Oppenheimer's quote from the Bhagavad Gita to ourselves: Have we become Death, the destroyer of our world?

Just more words

We have to stop carbon dioxide emissions entirely within the next thirty years. If we succeed, humanity may prevent the Earth from warming by more than 1.5 degrees Celsius, relative to its average temperature prior to the Industrial Revolution. What are the chances that humankind can agree on that one goal?

If one thing characterizes our time, it's the struggle over words, the power to define the world and its economy, the power to report and shape the news. That struggle is about deciding how the world is worded. Words create reality; owning words and the means to distribute them is crucial to all powers.

Shortly after interviewing the Dalai Lama, I was invited to a literary festival in China. In an airplane between two of the cities jointly hosting the festival, Chengdu and Beijing, I watched the movie *Social Network*, which is about Mark Zuckerberg, the founder of Facebook, and his desires and escapades. It was a peculiar feeling, watching a film about such a ubiquitous entity as Facebook while traveling in a state-owned airplane in a country where Facebook is banned, where phenomena such as status updates and thumbs-ups are considered a threat to national security. In one of the movie's key scenes, one of the friends behind Facebook writes an algorithm on a window in pen, finally discovering the "like" function as if he were coming up with the theory of relativity itself. Until then, I hadn't realized the "like" button was our generation's masterstroke. It has split our atoms by hacking the human ego and unleashing its previously unrealized energy.

In China, back in 2010, I remember I was surprised that Facebook was banned. Surely the website was a bastion of freedom? At the time, it hadn't dawned on me that Facebook was tricking me into filling in all possible information about my interests and political views. I'd allowed it to monitor my travel and my reading habits; the algorithm even listened to key words in my private chats. What's more, all this information was being sold to the highest bidder like some sort of buffet. My presence on Facebook undermined other modes of mass media and literary creativity itself. I was sharing news that made me happy or angry and the program was then reinforcing this behavior by showing me primarily things that buoyed these very same feelings. I'd started living in a world where the words that were topmost on my mind were getting amplified. I'd allowed unknown systems to analyze my behavior and thus I'd made it easier for companies to sell me things I didn't know I needed. And worse: to allow undisclosed parties to supplant my opinions.

In China, it's abundantly clear that people's views are being steered by larger systems—yet we risk overestimating the freedom we enjoy in our own contexts. When Iceland assumed the presidency of the Arctic Council, the U.S. fought against certain wording in a report about climate change impact. It is the official policy of the Trump government to deny such impacts and to remove words related to climate change from public records and websites. The U.S. Secretary of State, Mike Pompeo, talked about the melting of Arctic ice as "a new business opportunity." Commercial sailing routes to Asia could be shortened by up to twenty days.[56] On the one hand, climate change is denied while on the other hand, the business opportunities arising from the same change are cherished.

With the world's oil interests amounting to roughly $600 billion a day, oil producers have shaped our vocabulary and our worldview in line with their preferences. The oil industry has been fighting a propaganda war against climate scientists for more than thirty years in order to protect its interests. Oil companies have bought politicians and run "think tanks" to confuse the public using the same methods that were deployed to dispute the risks of smoking. Contradictory messages and false science are regularly distributed around the world—not without success. An anti-science president wields power in the United States. By sowing doubt and misconceptions about climate, the debate has been systematically confused and obstructed.[57]

People quarrel over words, over who gets to shape reality in mass media, and it's unclear whether the words that spark thought and form opinions serve the individual or speak on behalf of larger interests. Sometimes certain words get "turned off" so they cannot be heard. A recent report from the U.S. Energy Information Administration (EIA) outlines predictions for the U.S. energy industry up to 2050. The words "global warming" and "climate change" are nowhere to be found in the report, nor is there any discussion of what scientists concluded at the United Nations Intergovernmental Panel on Climate Change (IPCC), that emissions should be halted entirely by 2050. On the contrary, the car fleet is expected to be largely still run on gasoline by 2050 and the United States will become an exporting country for oil and gas by 2020.[58]

The phenomenon of global warming has been systematically removed from all official materials emanating from the White House; the same is true of most government websites and climate information sources in the U.S.[59] Powers with oil industry interests took advantage of Facebook and fake news to hobble the government, which should be the custodian of freedom and of information. We're heading at full speed over a cliff edge and instead of trying to brake, the gas pedal is floored and the speedometer's missing. As George Orwell put it: "Ignorance is strength."

The words the Dalai Lama spoke to me cannot be heard in China. Anyone in Tibet found with his picture or with any of his writings is in trouble. In Beijing, I met a person who talked about the Tiananmen Square events in 1989; he described the hope that had been kindled there and the bloodbath when the protests were beaten back. I talked to another person about the event. He hadn't been on the scene but said it all had to do with complaints about student loans. Students camped in the square, but there weren't any toilet facilities. The police came and asked people to leave—then people left.

"Did anyone die?" I asked.

"No, no one died."

I asked a Chinese girl about the Tibet situation. She explained it to me carefully. Tibetans were oppressed and subjugated and lived under a religious dictatorship. She asked would I like it, living in a country governed by a medieval pope or by clerics, like in Iran? I couldn't agree but decided against starting a debate over democracy. At that moment, our own democracy felt more than fragile. I knew I was the product of my European environment and upbringing.

My idea of China stemmed from "Oriental" fairy tales and legends and from Cold War propaganda, and even from old, ingrained Western Sinophobia. But it had also been shaped by multinational companies that abandoned human rights and environmental requirements in order to produce products as cheaply as possible.

The first thing I encountered at the Beijing airport was a black Range Rover; that unnerved me because these vehicles were symbolic of the financial crash in Iceland. China had been in a state of poverty just a few decades before; now there were gas-guzzling luxury cars on every street corner. The biggest building bubble in human history was in full swing, cranes everywhere surrounding high-rise buildings, all disappearing into a gray cloud of pollution.

All the same, I happily imagined spending a longer time in China. I found the country both pleasing and beautiful. In the parks, cherry blossoms flourished and old women did *qigong* while flocks of people practiced ballroom dancing. The Chinese people I met were comfortably informal, as are Icelanders.

We drove up to a mountain village in the Huidong area where villagers came out in the evening and danced a ring dance in the square; it reminded me a little of the Faeroe Islands during Saint Olaf's Wake. The village was fairly new. We visited a museum about the Yi people who'd written a kind of runic alphabet, enjoyed wrestling, and appeared to worship sheep. Again, quite like Icelanders! I visited a school and read to the children and then they read to me in unison; it was beautiful and they seemed happy. I was told most of them lived with their grandparents because their parents worked in factories in the cities. In the village, I saw children with new bicycles and bags of candy. The buzz of newfound prosperity hung in the air. Almost none of the children had siblings; it was striking to think this was an entire country where the words "brother" and "sister" barely existed.

We proceeded at a snail's pace through Chengdu and its throngs of traffic, passing along some of the biggest overpasses I've ever seen, massive structures through what was once one of the city's historic districts. I'd hoped China would think two hundred years ahead. But it seemed the Chinese were intent on repeating all the twentieth century's mistakes on an even larger scale, creating a disposable economy doomed to gorge on Earth's natural reserves and crocodile habitats. Towering skyscrapers seemed designed to serve contractors' interests rather than to shape a workable community for the future. At the airport, I met a man who worked for a company that made subway cars. I asked him about Chengdu. Why so much traffic and these enormous freeways instead of a vast, functioning subway system?

"The Chinese are human," he said. Contractors and large engineering firms no longer able to design and build overpasses in the West have targeted China, along with their allies, the ambassadors from the car manufacturers. Here's where the *opportunities* are. Western luxury manufacturers have seen a new market. Chengdu has more Lamborghinis than Europe.

China's building bubble absorbed 50 percent of the world's total raw material production in 2010. The Kárahnjúkar power plant was built in record time primarily to meet this demand, drowning out the all-encompassing silence of God's great expanse in Kringilsárani. China used more cement in each of the three years after 2004 than the United States used throughout the whole twentieth century, a statistic that's impossible to process.

We drove for miles, past half-constructed houses, block after block after block, forty-story towers standing on the ruins of old Beijing neighborhoods. The driver told us the apartments were empty. People invested in them, but did not want to rent them out: rental rates weren't high enough and a brand-new apartment was easier to resell than a "used" apartment. These were hundred-square-meter bank accounts. By 2018, around fifty million apartments stood empty in China, according to estimates.[60] Enough to accommodate the entire population of Germany—and France, too.

In Beijing, we drove through a neighborhood that was still under construction. On its outskirts, there was a Disney-style shopping center with a McDonald's and a Pizza Hut. The neighborhood's symbol was a child standing with a violin in its hand on a little jetty that jutted into a lake. The neighborhoods were called La Grande Villa, Palm Beach, and Shadow Lake. The dream of living in single-occupancy waterfront houses, in luxury Las Vegas- and Dubai-style villas amid forty-story high-rises.

The Shadow Lake villas, set around golf courses and lakes, seemed to be part of the area's branding, a dream of luxury designed to cast a positive sheen on the area, helping raise the value of the surrounding neighborhood. The problem with selling a house by a pond is that Beijing is one of the driest and most densely populated areas in China, with groundwater levels dropping by one to three meters annually and the world's largest water supply system bringing water to the city from 1,500 kilometers away. Other water pipes from Tibet are being planned.

It all seemed shortsighted. Many people had told me that the communist government demonstrated sensible, long-term thinking, but here it was hard to see any other force at work than greed. Contractors built high-rises as rapidly as possible so they could use their profits to buy hundred-million-dollar villas on man-made lakes.

The much more positive news is that nowhere else in the world are there as many science and engineering students as in China. Hopefully the nation and the Party will wake up before the situation becomes untenable. The Chinese are already at the forefront of solar energy production and wind power, and perhaps the world cannot be saved until the Chinese do something green on the scale of the Great Wall.

Planting a billion trees will take a moment if all of China works as one. One day, there could be sixty million wind turbines too many in the country.

Scientists have said that the global economy must change in a few short years in order to counter a two-, three-, and even four-degree Celsius rise in the Earth's temperature. Any hope that humanity can unite around a shared idea seems especially far off in Israel, where events that happened thousands of years ago and archaic manuscripts divide people into incompatible groups and alliances. I was in Tel Aviv during a documentary film festival where I met a boy and girl who showed me around the city. A large part of it was built in a modernist style by architects who had studied at the same time as Gunnlaugur Halldórsson, the architect who designed the workers' apartments Grandpa Árni moved into on Ásvallagata. During our conversation, it emerged that my tour guides were apolitical. As they uttered that word, five military helicopters crossed overhead, flying toward Palestine. I wonder what it means to live in such a place and claim not to have an opinion. Maybe this was the nearest thing they had to radicalism: being apolitical.

I went to Jerusalem and walked up Via Dolorosa. I tried to recall the Bible stories and found it a little too much, being on this holy ground. I felt as though I'd need to have my mind cleansed before walking down this street where Christ carried the cross up to Golgotha. I tried to think of the suffering of Christ as my eyes scanned the windows of the souvenir shops along the street.

Just off Via Dolorosa is the Wailing Wall, the only remaining wall from the temple destroyed in 60 CE. I had not realized that the Wailing Wall was in fact an old platform for the temple; where the temple stood is now the Temple Mount, one of the three most sacred sites for Muslims.

I knew all these places from newspapers and books, but I had not realized how much of a focal point they are. The surface of the Earth measures 510 million square kilometers and the holiest places for both Christians and Jews, and one of the three holiest sites for Muslims can all be found in the same square kilometer. That small piece of land is the religious center for four billion of Earth's inhabitants, most of them further subdivided into sects, many thinking they have nothing in common with the others. Everyone interprets words in their own way and so it has been for hundreds, even thousands of years. Even the Church on Calvary Hill is divided between five or six denominations, and there is a constant tension over territory.

Setting aside religion and politics, most of the human race seems to be wearing the same Nike shoes, with the same Samsung phones in their pockets and the same Sony TVs at home, with the same movie stars on their screens, using the same electric mixers, measuring the universe with the same units, playing the same music in their ears, rushing around in the same Airbus jets and Benz buses, munching pizzas and hummus over soccer matches played by global rules. Perhaps it's quite possible to shift our underlying energy system and our material reality while people continue to divide themselves into political parties, religious groups, and supporters of soccer teams.

The glaciers that my grandparents' travel companions once measured have allowed experts at many of the world's major universities to calculate what the future will look like. Natural scientists are starting to say things in very precise words: we must reduce greenhouse gas emissions, otherwise the world will be in trouble. At the same universities, professors of business, marketing, and engineering are steering world manufacturing in exactly the opposite direction. They forecast growth and increased consumption as positive signs. In one university building, people are figuring out the serious consequences that will result from endless growth on a finite planet while, in another building at the same institution, people are teaching limitless market growth, production, and waste.

The aluminum industry, one of the world's most energy-intensive industries, set itself the goal of tripling world production from 1998 to 2020. Even in 1998, everyone knew that the foundations of our ecosystems were beginning to erode. In 2005, when world aluminum production was thirty-two million tons annually, it was evident that humanity had hit the earth too hard.

How do people set a course from that point? Why not double something in fifteen years that already rained blows on Earth for one hundred? Such increases cover nature's immense sacrifices, including the loss of the area where Helgi Valtýsson found the all-encompassing silence of God's great expanse. The Alcoa smelter that drowned Kringilsárrani produces 340,000 tons of aluminum annually. The all-encompassing silence now accounts for 0.5 percent of the world's sixty million tons of aluminum production.

On average, the world's aluminum industry emits about eight tons of CO_2 per ton of aluminum produced, making about five hundred million tons annually, almost 2 percent of the world's emissions. The world's steel production has increased by one billion tons since 1990. About two tons of CO_2 get emitted per ton of steel. World output has grown in all areas in recent years. This also applies to the plastics industry, the paper industry, the fashion industry, the automotive industry, the energy market as a whole, the construction industry, and meat production. Fifty percent of the CO_2 in the atmosphere is due to emissions since 1990.[61]

Those who define the world based on money, industry, and production capacity have seemingly been spared from acquiring an understanding of biology, geology, or ecology. They calculate statistics and feel optimistic. What's fatal to the Earth and unsustainable for the future is hidden by the words "favorable economic outlook." Increased oil production is positive for the economy; doubling aluminum production is positive. Economic growth doesn't distinguish sustainability and unsustainability. Imagine making no distinction between strengthening or fattening, or between a child or a tumor growing in the womb. Growth is simply presented as an inherent good; there's no distinction made between malignant and benign growth.

In Iceland, increasing car imports is part of a favorable economic outlook. In 2014, a road was laid through the landscape of Jóhannes Kjarval's famous painting of Gálgahraun, following an estimate that traffic would increase from four thousand to twenty thousand cars per day. The forecast was taken for granted in this country where glaciers melt at the fastest rate in the world. Why should rising economies in China or India learn from the mistakes of the West when we've not ourselves adapted to the science that should underpin important decisions about the future? The ideal seems to be repeating mistakes on ever larger scales.

When two of my grandma's siblings, Gudrún and Valur, were born weak and vulnerable, a priest was called, not a doctor. And they died. Now that the Earth is in trouble, should we call an economist or an ecologist?

See the blue sea

To describe the berserking surf surpasses me.
I find that words fall to the ground as
helpless as tiny brokenwinged birds.
— Jón Trausti, 1911

When do you really know the sea? I live on a planet whose surface is 70 percent ocean and I live on an island surrounded by turbulent waters and yet I suspect I don't know the sea. Some of my ancestors were sailors—Grandpa Jón was part of the coast guard, and Grandpa Björn worked on a herring trawler—but I've never really been to sea. I spent two summers taking a sailing course and one summer, when I was fourteen, I worked crabbing: spider crabs were brought ashore alive in large pots and we boiled them and sliced off their legs to squish out brownish meat for market. We peeled the shell off the body, cleaned away the gills and innards under a water jet, and squished the belly into white gunk. I hooked a cod in Breidafjördur when staying with our friends on Flatey, an island in the west of Iceland. When we gutted the fish, its stomach turned out to be full of spider crabs. I was able to play the expert and explain that the males have a triangular apron but on the underneath of a female the apron is larger and more rounded, so as to make space for the fertilized eggs.

Iceland is 103,000 square kilometers in area; its exclusive economic zone is significantly larger, extending some 320 kilometers from land, a total area of about 758,000 square kilometers. Iceland can be said to be 87 percent ocean, based on this. I once read a book about Icelandic fish and it surprised me how few species I knew. In childhood, we learned the names of the major mountains, but the depths of the oceans were less relevant. The educational system in Iceland places little emphasis on marine life; it was probably designed according to a Danish model when Iceland's territorial waters only extended four nautical miles from the coast. In a TV quiz show pitting Icelandic college students against one another, a picture flashed up showing four fish: haddock, catfish, cod, and perch. The country's brightest kids couldn't identify these really common fish, despite the fact that tens of thousands of tons were being exported every year. Haddock arrives at our houses filleted, so children don't see its spot and its black stripe. They no longer hear the folktale about how the visible black spot and stripe on a haddock are said to come from the time the Devil tried to catch it and left his fingerprint and claw mark on its skin as it wriggled free.

If you think of all the fish that live in the depths around Iceland, the demonic deep-sea fish, the seals, the dolphins, the porpoises, the shoals of capelin, and the whales, the largest creatures that have ever lived on Earth: given all this, every Icelandic child should marvel over the sea. Every ten-year-old should dream of becoming a marine biologist because a large part of the depths is still unknown and uncharted. But we have been raised with a partial fear of the sea, even a resentment toward the whale, anxious it will "eat us out of house and home." Seals are shot as soon as they near estuaries in order to protect the interests of salmon fishers.

Where I live, I don't see the sea every day, but each summer we stay on Melrakkaslétta in the northeast of Iceland, where we own a little abandoned farm literally at the end of the world. It was soon clear to us how many threads connect a single shore to the whole rest of the world. Arctic terns nest next to driftwood from Siberia; mines from World War II lie about alongside rusted oil barrels marked *Wehrmacht* and a Russian listening buoy from the Cold War. The terns hatch from their eggs and fly to South Africa eight weeks later. On the shore one can find plastic bottles of Ajax cleaning detergent, brown bottles with an odd gunk at the bottom; a practically undamaged right shoe; and a Filipino sailor's safety helmet, labeled *R. Marquez* — that's how I learned that Spanish naming customs are prevalent in the Philippines.

A short distance from our farm, there's an inlet where the seals nurse their pups and bask on rocks. My daughters have sung some fun songs to them and sometimes the seals assemble, ten or so of them, for a concert. From Seal Cove, gannets can sometimes be seen drifting in from Raudanúpur. They are some of the most graceful creatures on Earth; they nest in a dense crowd on cliff-top crags. I've watched the gannets fish through my binoculars: they coordinate into a ring and stab down into the sea like fighter planes. Ornithologists have told me it's called plunge-diving and that gannets can dive to a depth of twenty meters.

Grandma Dísa used to collect eiderdown up on the gravel beach while we kids searched for treasures in the tide wrack. We once found a Puma-brand soccer ball with the name *Jonathan* and a phone number written on it. We dialed and the ball turned out to have gone out to sea at Sørvær, the northernmost town in northern Norway, 1,300 kilometers away. It had been on the way to us for a year and was in perfect condition so we mailed it back. I also found a message in a bottle lying right on the Arctic Circle at the farm at Rif. The message was from a man living in Oswestry, England. His name was Andrew, my name is Andri, so we met up a few years ago on St. Andrew's Day. In a movie, this would have been a beautiful love story; he is now my informal agent in the British Isles and has organized reading tours for me there. Flotsam has its benefits.

Perhaps it sounds like the shoreline is full of plastic waste, but as children we never saw the garbage as garbage or even as pollution; these were treasures we could collect and play with. In a strange way, everything that washed ashore became a kind of nature, more often than not crumpled and weathered and covered with barnacles and seaweed. The plastic at the shore is visible and messy and can cause considerable damage to the biota, with birds and fish tangled in nets; when plastic breaks down into microplastics, it can enter the food chain. But trash is first and foremost a symbol of our disrespect for Earth, how disconnected we are from the cycles of life. Refuse and dead animals have always nourished other species, but we are the first species to produce waste that is toxic, useless, and that, in fact, harms nature.

In the ocean, plastic gets tossed about and the motion of the current causes it to form plastic islands: the great Pacific garbage patch is twice the size of Iceland's exclusive economic zone, twice the size of France. The plastic is unkempt, although that in itself does not alter the ocean, does not change its temperature, acidity, and salinity, the strength of its currents or the frequency of hurricanes. Those are the areas where our effects are the most severe, and those effects are caused by burning carbon fuels, human volcanoes of burning oil that flow and blaze continuously, every day of every year. Those effects are not so visible as debris that stings your eyes. We can't fix that situation by "tidying up."

Lifestyles in the world's most prosperous areas have a greater impact on the acidity and temperature of the ocean than those in poorer regions. The situation is actually worst where everything seems neat and tidy on the surface. Carbon dioxide emissions are at their highest in such regions as is the amount of waste produced and sent to landfill or even exported to developing countries.

I've sailed the distance from Norway to Iceland and have experienced the vast expanse our ancestors crossed to get to Iceland over a thousand years ago. I looked out over the deck and despite being out on the open sea, to me the ocean was like a closed book, a choppy, frightening surface underneath which lay a mysterious three-thousand-meter depth. This was a memorable voyage in part because the ship belonged to Peace Boat, a Japanese humanitarian organization. On board, I met Masaki Hironoka, who described his memories from the time he was five years old. On August 6, 1945, he was out playing when a nuclear bomb exploded over Hiroshima and a brown wind blew him down. He roamed the ruins in search of his father and finally found him in the darkness of his home. He had been in a streetcar whose window exploded, melting into his back. Masaki's father asked him to pick the glass fragments out of his back with pliers but he didn't trust himself and hid until his father died. The now-old man cried when he told the story, but by telling it he wanted to prevent such a tragedy from recurring.

After two days of sailing, a Norwegian oil rig appeared on the horizon. It was surreal to see this structure out in the great ocean in all this vast expanse. It's difficult to understand how such a drilling platform could threaten the ocean itself, not from an oil spillage but just with its regular, everyday operation. It was pumping up the magic liquid that gives us the superpower by which we propel ships and the planes we could see drawing streaks over our heads. When oil burns, it forms carbon dioxide, around 30 percent of which has been absorbed by the oceans, leading to ocean acidification. (Recent estimates suggest that between 2002 and 2011, the global oceans absorbed 26 percent of carbon emissions.[62]) Around 29 percent is taken up by land, and the remainder, around 44 percent, hangs in the atmosphere and absorbs heat, that would otherwise be reflected into space, causing global warming. Carbon dioxide thus causes energy imbalances across the Earth. Warming and acidification are both direct threats to the ocean. The oceans absorb about 90 percent of the heat that accompanies global warming; scientific research shows that the warming of the oceans is on a scale equivalent to the denotation of four Hiroshima bombs *per second*.[63] The effects of this warming are already visible in coral reefs worldwide.

I've only once come into contact with a coral reef, the Tobago Cays coral reef off Union Island in the Caribbean Sea. We were traveling with *Activ of London*, a three-mast schooner I was invited to sail aboard from Trinidad; we would explore the Caribbean waters for two weeks. Captain Jonas Bergsöe is Danish and his dream is to use the ship to sail to distant places and create a community of scientists and artists. The ship is like a living being Jonas keeps alive by sailing it as much as possible, because as soon as it docks, it begins to fall apart.

There's some primitive nature in sailing ships: their cotton sails, their oak decks, their masts, which are basically three giant trees. We had to water the deck to keep it from cracking, the ropes each have a name and purpose, the result of many thousands of years of evolution and experience. It's no more than a hundred years since most of the world's shipping was powered by wind, which was how people explored, engaged in piracy, and traded slaves. As we traveled, dolphins frolicked at the ship's stern.

Originally, the plan was to go up the Orinoco River in Venezuela. I wanted to visit some scientists who had worked with Uncle John Thorbjarnarson saving crocodiles, but the political situation was dubious, the country on the brink of civil war, and pirates were attacking those who risked coming too close to shore. We therefore set a course north, through the islands and skerries belonging to Saint Vincent and the Grenadines: Carriacou, Petit St. Vincent, Petite Martinique, Union Island, and from there to Tobago Cays. On the way back, the course was set for Bequia and Tobago and eventually anchors were to be cast at Chacachacare, an abandoned island between Trinidad and Venezuela.

We reached Tobago Cays just before dark after a week of sailing. This is a so-called horseshoe reef around five uninhabited islands or sand reefs; the nearest settlement is on Union Island.

Earlier in the trip, we had come across a small white-sand reef; on its surface lay stacks of something resembling burned driftwood logs. These turned out to be coral logs as thick as a hand's width that had been swept up by some hurricane. The logs indicated that a mighty coral reef was there, or had been until recently, though I was not familiar enough with the local conditions to swim down to look for it.

I'd seen coral reefs in wildlife documentaries and for some reason I had thought of corals as soft and spongy creatures but they are more like barnacles or sea urchins. Corals aren't plants but rather cnidaria, which live in complex cohabitation and build their own structures out of calcium carbonate.

That coral branches break and wash ashore is part of the natural cycle. The coral reefs are constructed on the skeletons of their ancestors; corals wash ashore where they crumble and turn into snow-white sand and thus form the reefs and white islands that protect coastlines from ocean waves. In the logs we could clearly see how powerful the geological phenomena of corals and calcification are, but the cycle is based on the balance of death and regeneration. The increased intensity of hurricanes and the rising sea temperatures disturb the balance. Coral bleaching happens as these animals get stressed amid unfavorable environmental factors. The coral gets rid of the algae it has been living in symbiosis with, thereby losing its color, its branches becoming ghost white. Corals can recover but repeated neglect causes the animal to starve and die.

We cast anchor beside the coral reef and saw the waves breaking all around us at the reef's outer edges. The sea was clear blue and the bottom white. We floated toward the next island and encountered groups of strange fish. I didn't know what to expect given news about massive coral bleaching in recent years.

My concerns disappeared as we entered this otherworldly landscape. There were stretches of so-called brain coral, its surface and color like a human brain or a maze, its scientific name *Diploria labyrinthiformis*. They may be small, palm size, but many were wider than my arm span, greenish brown and round, like large moss tufts. The scenery that immediately came to mind was the mossy lava of Eldhraun in the south of Iceland. The brain coral reef stretched as far as the eye could perceive; I floated over a labyrinth of labyrinths like in some surreal dream. Inside there were staghorn corals and fans and I swam through a shoal of multicolored fish. Out of the corner of my eye I saw a stingray glide past, flapping its big wings, and there was a smallish shark and some silly clown fish. As it got deeper, we arrived at a white sandy seafloor with green seagrass. Giant white-gray crustaceans lay on the bottom at regular intervals, as if the mirror cells of life reflected the stars themselves.

I swam on and soon clapped eyes on giant green sea turtles (*Chelonia mydas*). In between floating up to breathe, they chewed seagrass in their jaws as though they were green cows. I hovered over them like a cloud in the sky. The turtles went gliding through the water, but then a school of fish with long snouts appeared, and I chased them through a maze of coral forms; they were multicolored and groping after food. I had no basis for judgment and did not know if the landscape was a shadow of something that had previously been more colorful and larger, but its beauty was still awe-inspiring. I emerged half hypnotized, sat on the shore, and freed myself from my flippers. A short distance away a coal-black tern sat on a tree branch and screeched. The turtle-cow grazed on seagrass on the seabed while a pink cloud in swim trunks floated above it.

Captain Jonas told us that when he was twenty, he had sailed alone across the Atlantic in a small sailboat. One night, he woke up to find a glowing fin whale following him a short way off, though we didn't fully believe his tale. After sunset, the sky was starry and we saw a strange light in the sea. I looked down to see that the fish moving to and fro had become luminous, as if they were radioactive. I drew a line in the ocean with a fishing rod and it was as if I had drawn a picture of the northern lights underwater. We leaped overboard into the darkness and a radiant, luminous coating formed around our bodies. This was phosphorescence, the *Noctiluca* bioluminescence, luminous single-celled organisms. When I raised my hand, my skin was covered with tiny luminous dots, thousands of them.

On the island of Petite Martinique, we met some men who were prizing shellfish from giant conches that had been thrown into a heap at the tide line.

"Where are you from, man?" the conch diver asked.

"Iceland," I answered.

He picked up a big white conch and handed it to me.

"Give this to Gylfi Sigurdsson!"

Soccer truly can unite us, I thought, if a conch diver knows about an Icelandic soccer player. The village on the island was dilapidated. On the other side of the channel were yachts and luxury villas. Petite Martinique's average annual salary would hardly have been enough for one night in the villa. The population had dwindled as people looked for opportunities on the bigger islands or moved to Grenada, the United Kingdom, or America. The main job involved fulfilling the dreams of rich people staying in paradise.

We landed in Tobago, the tropical forest island that is Trinidad's little sister, no more than three hundred square kilometers in size. Less than 5 percent of Vatnajökull's area. We entered a little bay, Englishman's Bay. There was a small shack and a crescent-shaped sandy beach, and thousands of pelicans sitting in the sea. Over their heads hovered frigate birds, coal black with their split tails, like a royal version of the arctic tern; indeed, they're known as "magnificent frigate birds," *Fregata magnificens*. They acrobatically dove down to the water's surface and picked up food that seemed to be there in large quantities. We waded carefully out in the sea, but the birds were tame. I put on a mask and then caught sight of a silvery-glittering strip of small fish that resembled sardines or capelin. The shoal formed a large circle, a gleaming whirl in the sea. I dove through the silvery whirl and raised my head right next to the pelican (*Pelecanus occidentalis*); it had a tremendously gray head and was so ancient in appearance that it seemed not to see me. I threw a fry at him, which he quickly scooped up. The pelican's face, like the wings of the frigate bird, seemed ancient, handed down from some long-forgotten avian ancestor. The sight was mesmerizing and timeless—fitting, given the pelican has been on the earth for forty million years.

Warm-water coral reefs are found widely throughout the world from the thirtieth latitude north to the thirtieth south. The world's coral reefs attract vastly diverse wildlife. About 25 percent of the biodiversity of the ocean is found there; they are sometimes called "rainforests of the ocean." Their total surface area is small compared to the size of the ocean; they're found within just 0.1 percent of the ocean's area. The largest coral reef in the world, the Great Barrier Reef, is a chain of thousands of coral reefs extending over fourteen degrees of latitude, more than 2,300 kilometers long and up to 250 kilometers wide off the coast of Queensland, Australia. It is one of the major gems of the oceans and home to some of the greatest biodiversity that the Earth has produced. According to the Australian Institute of Marine Science, the Great Barrier Reef in Australia suffered massive, back-to-back bleaching events in 2016 and 2017.[64] These events caused large areas of the reef to be depleted or even killed off.[65] Its regrowth has been slow; coral reproduction has decreased by 89 percent in many parts of the Reef since 2016.[66] A third mass bleaching situation occurred in 2020. This situation recurs around the world.

Cold-water coral reefs also exist but are less well-known. We usually associate corals with palm trees, but off the southern coast of Iceland one can find cold water and deep-sea coral. Biologists who have studied these coral reefs compare their diversity to entering an orchard from a desert. Mid-twentieth-century trawler fishermen would describe getting giant branches in their trawling nets, as if there was an underwater forest below them; the most extensive reef area was called the Rose Garden. Unfortunately, the majority of the Icelandic coral reefs were destroyed by the use of ill-advised fishing equipment in the twentieth century—symbolic of a time when technology and humanity's power had advanced far beyond our knowledge of nature.

Coral reefs are sensitive to harsh treatment, to overfishing and to runoff from cities, agriculture, and landfill. They can be harmed by oil pollution and chemicals in sunscreen. You get heatwaves in the ocean just as you do on land and these

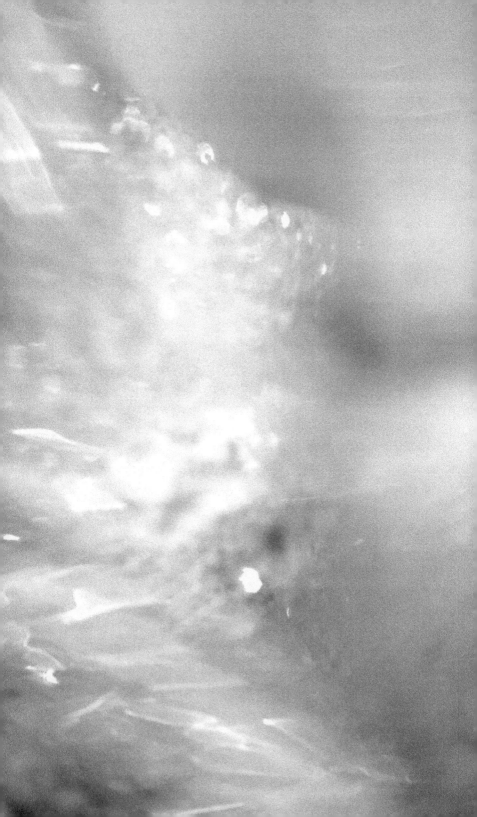

heatwaves have caused enormous damage in a short period of time, just as forest fires and the increasing frequency of hurricanes have wrought havoc. Rising temperatures, decreasing pH levels, and reduced oxygen test the resilience level of coral reefs and today all these forces work simultaneously. Just like the glaciers, coral reefs have over the centuries grown larger in some places and died out in others, but rarely have all the world's coral reefs had to fight for their existence at the same time. It is estimated that 99 percent of Japan's largest coral reef, located off Okinawa, has wilted, so only 1 percent of the reef remains healthy.

Our most important information about atmospheric CO_2 concentrations are the Keeling measurements made at Mauna Loa in Hawaii, extending back to 1958 without interruption, visualized as the Keeling Curve. One of the longest sequences of uninterrupted measurements of ocean acidity are those taken by Jón Ólafsson, a marine scientist in the deep north of Iceland; they date from 1984. They have not yet been named the Ólafsson Curve, but they show clear and rapid changes in ocean acidity and calcium saturation.[67]

For a long time, the ocean absorbing carbon dioxide was viewed as a buffer to atmospheric warming: the oceans take in about 30 percent of the CO_2 we emit. But we have realized that carbon dioxide does not disappear in water any more than in the air. Increased carbon dioxide causes ocean acidification; already, ocean pH has dropped by an average of 0.1 pH in the last thirty years, an indication of its increasing acidity.

And this brings us back to language: pH is a logarithmic measure, but we are used to thinking in linear terms: miles, grams, years, degrees. A logarithmic scale, where each unit increases exponentially, by a tenth, suits us poorly here. If milk was measured on a logarithmic scale, let's call it MOO, one liter of milk would be 1MOO, while ten liters of milk would be 2MOO and a hundred liters of milk would be 3MOO. With this scale one could easily get confused and order 3MOO of milk (a hundred liters) and not 1.3MOO (two liters). The logarithmic scale is well suited to those with mathematical brains, but for ordinary people it is especially impractical.

When warnings are made about changes in ocean acidity, from 8.1 pH to 7.8 pH, we don't recognize the extent of the difference: 0.3 is plainly a small amount when it comes to currency, percentages, meters, and years; even when considering millions of dollars, 0.3 is not a huge amount. In fact, 0.3 is usually within the margin of error in all our calculations. A child with a 0.3 rise in temperature still goes to school that day; 0.3 gets rounded to zero. A logarithmic scale used in an important matter like this can act like a language without adjectives. For example, human blood has a certain acidity that can withstand fluctuations from 7.35 to 7.45 pH. At higher and lower limits, a person becomes seriously ill. If the value goes beyond this limit, there is every chance of organ failure and death. For many animal species, the acidity of the ocean is as important to them as the acidity of blood in our body is to us. In fact, the underlying change in these 0.3 pH levels is so severe that the descriptive words need to be both uppercase and bold and accompanied by twenty emojis. And still the figure 0.3 is just one more black hole.

One of the consequences of acidification is that the calcium saturation of the sea decreases and the seawater becomes subsaturated. In simplified terms, it can be said that supersaturated seawater is rich in calcium carbonate, also called limestone; that raw material is therefore available to organisms that want to use lime, or aragonite, the name of the building material that provides the shells of most shellfish.[68] Supersaturated sea thus seeks to shed lime; subsaturated sea absorbs it, dissolving shells and coral reefs. One can talk about fundamental changes in the nature of the oceans; these effects are greater in colder seas than in warmer. Acid levels affect animals such as Thecosomata, sea butterflies, which provide close to 40 percent of the Pacific salmon's diet.

There was a beautifully written article in the *Viking Seaman's Magazine* about this small winged animal and its role in Icelandic waters:

Hardly anyone can imagine snails and conchs traveling around the ocean. The so-called sea butterflies, however, are snails that spend their lives like plankton, flittering very rapidly about on a type of wings so as not to sink to the bottom. Most of these animals are so small that humans are hardly aware of them, if at all. The calcium carbonate shells they carry give them a structure and beauty that is no lesser than what is typical of larger, floor-dwelling snails. However, their lime shells are paper-thin, almost transparent, lest they burden their owners. Older fishermen may be familiar with the brown gunk sometimes found inside herring. What they are seeing is shelled sea butterflies: the thin snail shells shatter to smithereens in the herrings' bellies. This suggests that the sea butterfly population can be enormous, but otherwise does not particularly stand out in the diverse group we call plankton.[69]

Experiments on sea butterfly snails have shown that those that live and grow in acidic oceans have a much thinner shell than those living in healthier conditions. They therefore have weaker defenses, are more vulnerable to impact, and spend more energy building and maintaining their shell. The same is true of many other calcium carbonate-forming organisms in the ocean.

If acidification and warming disrupt the life cycle of copepod species like the red plankton, *Calanus finmarchicus*, the ocean's nutritional cycle could entirely collapse. Red plankton are like a floating dietary supplement for the ocean; together with other zooplankton and phytoplankton, they form the basis of the marine ecosystem in northern oceans.

Planktonic algae produce 60 percent of all Earth's oxygen through photosynthesis and will be adversely affected by global warming and ocean acidification. Adult fish and even adult krill appear to tolerate changes in acidity reasonably well, but at larval levels marine animals can be sensitive to temperature, salinity, acidity, and calcium saturation. If all these factors go haywire, there is likely to be a collapse of the species that forms the basis of the food chain.

Ocean acidification is one of the largest unique geological events that the Earth has undergone in the last fifty million years. And it introduces another concept to which we connect poorly: *Time* itself. Although time is properly called linear, imagining that the ocean will change more in the next hundred years than it has in the last fifty million years is a challenge. The time since Iceland's settlement is very short, not more than twelve times my grandma's life: eleven hundred years. The history of Iceland is, in a sense, a continuous story of twelve women like my grandma. Twelve girls who were born and lived lives that each felt like a flash. Twelve women in their nineties stretching out their hands as if they're doing water aerobics, touching flat palms together. Their eyes gleam because time passes so fast that their eyes don't realize they're nearly a hundred years old. Time runs so fast that Jesus was born around twenty-one grandmas ago. They'd all fit in a single city bus, even if you added all their husbands. The earliest written records of humans date back five thousand years, events that happened practically yesterday. Humanity first emerged the day before that, in comparison to the ocean's fifty-million-year history.

The term "ocean acidification" had appeared three times in print in Iceland in 2009, a frequency more or less matched in many other countries. When Jørgen gave us freedom, it took us a hundred years to fully understand that word, but we needed to understand acidification thirty years ago. The consequences if the oceans reach 7.7 pH will be more frightening than words can say.[70] People can cast doubt on climate change and point to small and large ice ages, but not so for ocean acidification. The process happening right in front of us has no parallel to those we see with ice age fluctuations.

I had high hopes for the Paris Agreement and for international climate talks. I reckoned the agreement between the world's principal leaders to prevent a more than two-degree-Celsius rise in Earth's average temperature would keep us inside the limits of risk. Everything beyond that would be a danger zone. But then I met scholars who told me that wasn't the case. That at a two-degree rise, all kinds of disasters would already have taken place. I met the scholar Joni Seager, who talked about "climate horrors": desertification and water scarcity and associated wars. I was shocked when I heard this because in my simplicity, I believed that we could "stick to" two degrees. And the two-degree goal is not even in sight; the world is heading toward a three- or even four-degree rise. The living world is subject to uncertainty; scientists are attempting to figure out the most complex systems imaginable. We are least certain about the specific tipping points. Think of snow cover on a mountain slope, which holds in place until the exact snowflakes that disturb its balance land, causing an avalanche to come rushing down the slope. The mass holds until the precise tipping point; then it fails, and everything alters.

Water is a good example of a material with clear tipping points. When the temperature of ice rises from minus fifty to minus ten degrees Celsius, nothing seems to happen. Nor even when ice reaches minus five degrees or minus one degree, even though the temperature has been changed by forty-nine degrees. But just one more degree changes everything: ice becomes water. Antarctica and Greenland are home to the largest ice sheets in the world. We do not know how far off their tipping point we are. Projections of sea level rise could skew by tens of meters. The acidity of the ocean could drop by 0.1, even 0.2 pH without people being able to measure large effects. But add an additional 0.001 value, and there may be irreversible failure.

Another tipping point that causes uncertainty is the permafrost in Alaska, Canada, and Siberia. If soil that has been frozen for thousands of years thaws, then microorganisms will come to life that emit nitrous oxide—laughing gas—a greenhouse gas three hundred times more powerful than CO_2. Methane, twenty-five times more powerful a greenhouse gas than CO_2, will also be vented. Simultaneously, we'll see the oxidation of the once-frozen soil. As much CO_2 will discharge as when the wetlands have dried up. These emissions will warm the Earth and release yet more methane and nitrous oxide. If such a process is initiated, all our discussions about reducing our carbon footprint through diet and via *flygskam*, "flight shame," will seem completely ridiculous. Such a chain reaction could throw the Earth into a period of complete atmospheric and climatic chaos. The only remaining plan would be to hoard dried food and fasten our seat belts.

The 2018 UN Climate Change Conference in Katowice, Poland issued an emergency call. David Attenborough came onstage and was unusually harsh in his words:

> If we don't take action, the collapse of our civilizations and the extinction of much of the natural world is on the horizon.[71]

For most, it was more buzz, and yet it was still as though understanding, somehow, had begun to take shape. The conference message was clear. Humankind has to see to it that Earth's temperature does not rise by more than 1.5 degrees Celsius. A two-degree Celsius increase would mean sacrifices that no one could accept.

In connection with the climate conference, I watched a gathering of scientists streamed online. It would be fair to describe the speakers as measured. Their emotions were in check; I thought again of mass apathy vs. mass hysteria. A marine biologist came onstage and was particularly upbeat. He said we had to meet the 1.5-degree target, because that way *only* seventy to 90 percent of the coral would disappear, and not all, as the two-degree Celsius rise projected. He spoke like this was a worthy fight. Despite my interest in these matters, I hadn't been aware of this. Was he telling me, so very directly, that up to 90 percent of the world's coral would die out if we reached an almost impossible goal of keeping global warming within 1.5 degrees Celsius? Was that knowledge available when people were deciding whether to aim for the two-degree goal? Was there a group of councilors who approved this on behalf of the Earth's inhabitants? I thought back through the news from recent years and couldn't remember television and radio broadcasts being interrupted by this decision. I do not remember an election where a nation mandated the elected government to sacrifice the world's coral reefs. I didn't recall coral reefs being used as leverage in negotiations: "The car manufacturers' association celebrates victory. The coral reefs had their time."

Why didn't this become an "event"? I remember where I was when the Twin Towers fell and when Princess Diana died; my mom remembers where she was when Kennedy was shot. Where were you when the coral reef death sentence was signed?

> I was at a gas station, planning to buy a foot-long sub and it was weird, the customers were standing about like they were paralyzed. The TV behind the counter was showing these odd fish . . .
> "What's going on?" I asked.

"It's over," the clerk told me, tears in her eyes. "The coral reefs lost."

The marine biologist evidently knew coral better than anyone else, but he wasn't pulling his hair out and running screaming along the corridors: "ARE YOU ALL ASLEEP? DON'T YOU UNDERSTAND WHAT'S HAPPENING?"

We're the generation that decided to sacrifice the coral reefs, and they're but 1 percent of the whole situation. The ocean produces about 60 percent of the Earth's oxygen when plankton in the sea's upper layers photosynthesize. No one knows where their tipping point is; getting to that point is a risk no one who lives on Earth can take.

In the fall of 2018, I gave a lecture about time and the ocean, and I was planning to show a video of the turtles in Tobago Cays and share some findings from marine biology about coral death. I'd written that perhaps this meant I was saying farewell to this turtle. Its thirty-five-million-year history was ending because humanity had made the decision that brands and businesses were more important than turtles and coral reefs. Hulda Filippía, my ten-year-old daughter, was in the audience and all of a sudden I felt so unbelievably sad. I didn't want to extinguish this light in her eyes. I simply couldn't show the video, couldn't reveal what lay ahead for this beautiful species and for its environment. I didn't want to tear up in front of an entire hall of people.

Future generations will ask about the values and priorities of the unfortunate generation that steered Earth when the decision was made to sacrifice all the coral reefs. They will look at combustion data and find that the richest nations in the world didn't feel a need to impose emergency laws regarding waste, combustion, and squandering. The average car in the West consumes two to three times more energy than it should, compared to the best-designed cars—and ten times more than well-designed public transport, and a hundred times more than bicycles or electric scooters. The average refrigerator consumes three times more energy that it should compared to newer, better-designed refrigerators. Production of beefsteak causes ten to thirty times more emissions than production of plant-based foods. They will see these priorities and judge us for them. Short-haul flights between major cities weren't restricted, even after we noticed their impact. No Marshall Plan for building wind turbines and solar farms was implemented; the world's engineers were not set challenges like those put in place to send humans to the moon. In fact, there was no restriction on people's freedom to pollute or release emissions in any way. Nature has almost no rights. Unspoiled nature gets viewed as under-utilized raw material. There are no plans to establish legal ramifications for crimes against the planet—for what is called "ecocide," equivalent to genocide. Ecocide laws would see the prosecution of those who had a role in the destruction of ecosystems; there is currently no accepted international crime of ecocide.

Debates about ideology, arguments between left and right, liberalism and conservatism will persist. No ideology or law, however, should sanction one generation's actions when they cause immeasurable harm to future generations, depriving them of so much that is of value. We expect functioning governments to restrict an individual's freedom to cause others harm; it is a fallacy of democracy if the system does not allow us to think decades into the future in this respect. Business interests and human comfort have been seen as more important than the ocean, the atmosphere, and all the world's grandchildren—for all time.

In Asgard, the gods made a deal in which the giants would build them a city. Then the gods refused to pay. But they paid a price when the frost giants took action. Today, the world's glaciers are melting. Enchanted frost giants are set free and plunge into the Arctic oceans and stream foaming white from the mountains, transforming into waves that crash into city gates.

If we want to save the coral reefs, we must rewind. Experts have deduced that coral reefs thrive in a climate below 350 ppm of CO_2.[72] We have far exceeded this limit, have reached 415 ppm. Based on that, we need to immediately brake and return all emissions to their levels from twenty-five years ago. If we are sensible creatures and see we are losing the crown jewels of the oceans, surely we'll take action, right? And if the frozen cows of the mountains, and the kings of the animals, and the world's agricultural lands are at stake, too? What will we do? Do we require more proof?

Do we still have no idea?

Maybe everything will be all right

From the age of three until I was nine, I lived in the U.S., catching frogs and tadpoles in a pond behind our house. Then we moved back to Iceland and I was suddenly at Melrakkaslétta, in the northernmost part of the country, where Grandma Dísa and Grandpa Jón spent their summers in a deserted farm three kilometers below the Arctic Circle. The house had no electricity and we had to chop driftwood to boil the water we'd collected in a bucket from a coastal spring. There, I experienced a quite specific sensation, one that was difficult to put into words, some kind of cultural shock. All the life there on the beach became almost like a provocation: flocks of arctic terns, the teeming common eiders, black-backed gulls, the nestlings that ran around the moor, the seals that stuck their heads out of the bay. I realized that it was not life that affected me the most, but death. Even on a short walk there was death at each footfall. The breastbone of a gull with its wings still attached. A dead catfish with a face like a monster. A sheep flat on its back, eyes full of wriggling maggots. Skeletons everywhere. Birds' beaks, crabs' claws, the whole coast covered with rotting seaweed. Black-backed gulls hovered over eider nests, dipping down and reappearing with wriggling baby birds in their beaks. Great skuas drowned eider ducks out in the bay beyond the range of Grandpa's rifle. We would find half-dead fledglings and try to save them but they would die in our hands and end up in a small burial ground behind the house.

In the city, there's no death. In zoos, all the animals are alive; death is invisible unless a snake is swallowing a rabbit or rat. In the parks, everything is neatly arranged and even on farms there aren't any carcasses or skeletons: animals exist inside their fences, in stalls and sties, all well sorted and even sprucely washed. In the grocery stores, there are aisles and aisles of products containing meat but there's no connection to death; it's often difficult even to see what kind of animal is involved.

The first time I traveled across the highlands of Iceland as a child, north across the vast sands by car, everything seemed to be lifeless and dangerous. We drove for hours at a time through wastelands and deserts, an area of mysterious, melancholy place names: Víti (Hell); Daudagil (Death Gully); Ódádahraun (the Lava Fields of Ill Deeds). At first glance, everything was gray and dead, but here it was not death but life that mattered most. The earth that seemed so gray was, on closer inspection, actually a flower garden. Between all the stones were tiny flowers, as if they had been placed there by a meticulous gardener. How could flowers be fertilized, root, and flower in such a place? From a puddle grew a single pricking blade of grass; in the melting snowdrifts one could see that beneath a thin sheen of ice neon-green mosses had sprouted.

In geothermal areas there were boiling hot springs; at the bottom and along their edges you could see a mysterious slime, a film above the bubbling clay pits that resembled bacterial flora, like the first organisms in the Earth's elemental soup. Areas that struck me at first as symbols of death and destruction became the best examples of the evolution of life. In the jungle, life is all-encompassing and self-evident, but here in the highlands it was naked and exposed, every blade a kind of miracle.

The area that I thought was the most beautiful of all was Askja. The lake there has the deepest water in Iceland; it's as good as brand-new, formed from a volcanic explosion in 1875, which makes it a contemporary of New York City's Brooklyn Bridge. The world we can discern is the remnants of places that previously had an entirely different appearance. Probably the endless desert of Sprengisandur was overgrown during the Age of Settlement and the Sprengisandur that fascinates me today would be a sad sight for someone who knew it a thousand years ago. On the plain at Melrakkaslétta, all the beaches used to be full of great auks and walruses. All gone today. Undoubtedly, the landscape will be endearing to someone in the future, even if it changes again. People live in the world of their own moment. They get used to their environment and cannot waste emotions on grieving over everything that has changed. We have the ability to see beauty in almost every circumstance; there are people who live in deserts and find beauty and depth in the emptiness—as do people manage who live at the poles, places that have hardly seen trees or flowers.

About twenty thousand years ago, Ice Age glaciers lay all over Iceland. The country slept for tens of thousands of years under two kilometers of thick ice that volcanic eruptions seldom breached. Under the glaciers there were mountains and valleys, but the ruthless ice had wiped out a unique landscape that had covered the land long before humans came into the story.

Before the Ice Age began, giant pines and deerwoods existed in the Westfjords; there were walruses in the fjords, otters swimming in rivers, woodpeckers sitting in the tree-tops, and perhaps crickets chirping on a late evening. Then the Ice Age glacier formed across Scandinavia, most of Northern Europe, all of Canada, and a large swath of the United States. In Central Park in New York, one can discover a rock hill shaped by a glacier, a *roche moutonnée*. White death once lay across these countries like a heavy burden; everything was erased. Today little that's alive in Scandinavia has been there for more than thirteen thousand years. Eleven thousand years ago, the Ice Age glacier was larger than Iceland; you can find ancient terminal moraines in the middle of the bay of Breidafjördur, many miles from where the coastline is today. Ice ages are like tides but on a larger scale. The glaciers retreat, allowing humans to settle and build cities; then the Earth's axis tilts farther and its orbit around the sun changes in eccentricity, shortening summers in the northern hemisphere and letting ice creep forward again. It takes more than one ice age to carve out a beautiful fjord; it took ice age after ice age after ice age to create the Norwegian, Icelandic, and Greenlandic fjords. Over the last million years, ice ages have come and gone in hundred-thousand-year cycles, building up for ninety thousand years then melting fast, followed by a ten-thousand-year interglacial period. Then the cycle begins again, like drawing breath: like seasonal fluctuations, but on a geologic scale. A thousand years of summers, then a thousand years of winters. We see the Earth revealed as though we were sandpipers rushing out onto the mudflats to scrabble up food before water floods the flats again. We migrate to our summer camps, then have to move back south when winter arrives. We make nests and have several generations of descendants, but then the ice-white eraser comes and destroys everything without mercy. An article in *Nature* in 2016 suggests that humans might have affected the planet enough to delay the start of the next ice age by over one hundred thousand years.[73] We are powerful enough to alter glacial cycles.

A glacier distends and thickens, devours everything without conscience, without economic growth, industry, greed, environmental assessments, or international conferences. A glacier scrapes everything away, like an artist who destroys his work as soon as it is fully formed. Where were the arctic terns fifteen thousand years ago? Were they on the coasts of Normandy or Spain? Where were the gannets and blackbirds? Where were the seals and walruses? How long does it take for a bird to develop and become an arctic tern that flies from South Africa north to the Arctic? It would not have been possible fifteen thousand years ago. There was nothing to come north for here, just continuous ice from Iceland to the pole. When was this a paradise for living things? When did the perfect situation exist, with animals on earth living in peace and equilibrium?

Fifteen thousand years ago, the sea level was a hundred twenty meters lower than it is now. Back then, the United Kingdom was not an island but part of Europe, and woolly mammoths and woolly rhinos roamed close to the edge of the ice, cave bears and Ice Age lions hid in caves, and people were scattered about here and there. Probably humans started eradicating species then and there, although few species would have been able to tolerate the alterations in vegetation and the rising temperatures that were to come.

There is no such thing as a permanent landscape; nature has no constant. Change is its essence. If it weren't for weather systems and volcanic activity or the moon that guides the tides, the Earth would be dead, or at best a stinking ball of algae. Nature is like the Hindu goddess Kali, who destroys as she soon as she births. She makes love as she kills, because creation and destruction take place simultaneously; in nature, there's no separation between them. It doesn't matter where we rewind back to in time, nature is always right, has always been true and right. Creation is change. Everything is in the process of transforming. In nature, the waterfall gradually forms deeper ravines or rapids and the glacier either retreats or else wipes out entire continents; tectonic plates push against each other and press mountains high into the heavens as other continents get swallowed and disappear into glowing magma.

Of all animal species that have lived on Earth, 99.99 percent have died out. One by one, they disappear for natural reasons, go extinct through competition, or end up in an evolutionary dead end. It can take ten thousand years for a species to fade out, though sometimes a faster end is caused by large-scale events, by giant volcanic eruptions or asteroid collisions. It is estimated that five such extinction cycles have occurred, around every hundred million years, and we have now initiated the sixth.

Europe is packed thick with human populations; most of the wildlife that was there has vanished. The population on Earth is headed toward ten billion and all that humanity can make use of, it will. There are still herds of wild animals in Africa, but when the world's population reaches ten billion, there won't be room for endless grazing space for gazelles, giraffes, and lions, just like there's no longer room for millions of buffalo on America's plains or roaming deer herds and wolves in Germany. Instead, we will see fields shaped into squares, if not for food, then to create ethanol or to pasture livestock. Wild animals will live in tight confines, in zoos, in small protected zones, in test tubes and seed banks. Wild nature will go the way of the wolves in Norway, bears in Germany, tigers in Nepal, and the Icelandic eagle. Humans are emotional beings; it will be painful to watch the sixth extinction happen in real time on-screen in wildlife documentaries and videos. Daily, we see news of the decline in the populations of rhinos, elephants, giraffes, and zebras. The Living Planet Index shows that between 1970 and 2014 there was an overall decline of 60 percent in population sizes.[74] The figure might increase to 80 percent in 2060 and 95 percent by the year 2080. The chicken has become the world's most abundant vertebrate. About sixty five billion chickens are consumed annually, and at any one time the combined mass of chickens on the Earth exceeds the mass of all other birds on the planet.[75] More than all the golden plovers, albatrosses, parrots, and penguins that disappear slowly when their habitats are encroached on and their nesting sites are lost. And yet the total number of birds on the planet has not decreased; it has increased. We've made up the numbers with more chickens. Kentucky Fried Planet.

Perhaps it will be necessary to let go and to take shelter in irony and apathy, to find ourselves a philosophy that reconciles all these shifts. If we are an Ice Age, we need to connect to the cold and live symbiotically with its expansion. We need to stop dwelling among the animals of the past and to celebrate the manifestation of who we are: the children of the black sun. We are the mathematical result of unexpected energy entering into the equation, a result that is asphalt, plastic strata, trash mountains, the chicken explosion. All these are traces of a species of primate that opened carbon veins in the Earth's crust and flourished like an algae bloom for hundreds of years before disappearing under grass and new sedimentary layers. For each species that disappears, we have a new brand ready. Tiger™. Apple™. Amazon™. We wait in lines and tents outside stores when new Nike shoes come on the market. Just as the romantic poets heralded the golden plover. And sneakers are nature, of course, part of the human Ice Age. Humankind is a species of animal. Our products are nature, my computer is nature, like a beehive or a Stone Age ax evolved by whetting its edges with ever more accuracy over sixty thousand years.

That's how we can push the world away and escape the anxiety that comes with worrying about the Earth's future. Everything that has lived will die. Beauty can be found anywhere, even where the most horrific acts of violence have taken place. We sit on the moss tomb at Eldhraun without thinking of the disastrous destruction of the eighteenth-century Laki volcanic eruption. We drive by America's endless acres of corn where before there were indigenous peoples and roaming buffalo. We have pleasant days out at the beach in Normandy and stroll through the streets of Hamburg, where firestorms roared after the air strikes of 1943. Nowhere do we sense the weight of history; no sorrow can be seen in the eyes of the passersby. Europe recovered quickly after the war and in 1960 you couldn't see that fifty million people had died, that all the survivors had lost someone, killed someone, or fled their homelands. And maybe everything will be okay in the end, no matter what happens. We don't miss the ancient hunting grounds the North Sea conceals. Once the world's water levels have risen, people in the future won't be able to miss what's beneath that water. Water smooths over grief. Everything dies and is born simultaneously, forever.

Where the glacier stood out against the sky, beauty still prevails. The world does not have to be ugly as it changes, as we multiply or dwindle. I feel that if I calm my mind, then everything will be all right in the end, even if things go badly from this point on. I bring my mind in this direction hesitantly, flirting a little with nihilism, with apathy. The hypnotic thought that everything is relative. A seductive thought, like a siren song; it's tempting to let go, the universe is billions of light-years in scale and we're just a small spark in the context of those billions, we're really nothing but one more layer of ash in the history of Earth:

Nothing gets saved; things pass, things pass.
They crumple in on themselves and no longer exist.
Your life makes the smallest mark, a minor profit,
and then, finally, it is over. As if nothing ever happened.

These are Steinn Steinarr's words; this is one way to look at the bigger picture. I am grass and I will grow over your footprints. Why vote, why wake up in the morning, why shower or head to the gym, why write poems, love anyone, have children? All love fades; everything dies in the end.

Languages die out, books go moldy, songs get forgotten, works of art warp, anything we produce eventually turns to trash. All of this is as inevitable as the sun becoming a red giant and swallowing everything.

One of the last recordings in the extensive audio archive at the Árni Magnússon Institute was made the night before I was born: July 13, 1973. The recording contains environmental sounds from Básar in Grímsey, an island north of Iceland. You can hear foghorns, arctic terns, the waves lapping. The recording is strangely fascinating. Helga Jóhannsdóttir and Jón Samsonarson traveled around the country looking for the last individuals who'd grown up singing folk songs and knowing the oral traditions that dated back many centuries. But in that spot they preserved three minutes and fifteen seconds of a summer night in Grímsey. A minute snapshot of eternity.

I always wondered about this recording. Was it an attempt to capture beauty, or were they so engrossed in collecting that they wanted to preserve everything, including the eternally lapping waves in Grímsey? Was this an effort to preserve everything in the world? Or was the recording some kind of surrender?

Helga and Jón traveled all over the country to save what was left of a disappearing culture. Often, they found that the elder who knew the song, the poem, or the old legend was recently deceased, senile or voiceless; inestimable resources were thus lost to eternity. I imagined that July night was an instance of a kind of harmony with the world and the flow of time. Everything flows by, everything is eternal and transitory, like the Grímsey waves. The attempt to preserve anything is futile.

I could get used to this philosophy. Within it, I can find a tiny fragment of shelter from all the terrifying information, a bit of space to hold on to some sense. It's possible to let go and float downstream. But I'm afraid that future generations will despise that position. Someone who is twenty today could know and love another human who will still be alive in the year 2160. And, given the current shape of the world, we're heading toward such a degree of destruction that this emerging generation will judge all our lives as laughable and foolish. We'll be regarded as primitive and naive, just as people in Jørgen's time did not understand the word "freedom" and so let a small, unarmed group of monopolistic merchants keep a whole nation under tight rein. Philosophical surrender would be yet another symbol of our unruly egocentrism. And just like art that pleases dictators, all our creations will be valued in the light of the devastation we know our way of life has caused. The beauty of coral reefs and jungles, the dignity of extinct species, these will be lined up next to all the junk we craved and used only briefly before it all ended up in a landfill. Brain coral will be measured against traffic jams, the clothes we throw in the trash every day, the food we waste, the discarded soda cans that go on trash mountains, and the oil we burn for fun. If we do nothing, we will be the generation that was handed paradise and ruined it. And all because we stuck to our self-interest and our greed. All our work will be a crying shame because nothing we produce can hold a candle to the ocean itself, nothing is as magnificent as a glacier, nothing so mysterious as a rainforest at night. Nothing we have done will be considered remarkable if achieving it has meant closing our eyes to science and throwing away the life and happiness of future generations.

How to get around these figures? There's a gigantic river of oil, 666 erupting volcanoes, a hundred million barrels of oil a day, hundreds of millions of cars streaming along assembly lines and chugging along the streets like a river of lava. The number of cars produced annually could, set end to end, circle the earth four times, wrapped around it like a boa constrictor. Shot into space, they would form two Rings of Saturn around Earth at a distance of one thousand kilometers away. Looking up to the sky and seeing a shining circle of cars would be impressive, a sign of our power, before they come crashing to the ground like a meteor shower. Humankind has never before numbered seven billion. We have never lit so many fires at the same time. And now we need to think and behave differently than we did before. We have all the tools, all the devices, and all the knowledge to do it. And if we do not, we will fail both our ancestors and our descendants. The words are beginning to spin down into the black hole. Time to visit the holy man.

Interview with the Dalai Lama, in his guest room, Dharamsala

June 9, 2010

We fly to Delhi and stay two days. I've never been to India before, never experienced the way of life there. I'm not used to such huge throngs of people. I'm not used to the heat, I find myself bewildered and mistrustful, not knowing whether people are trying to help me, confuse me, or trick me—or all three at once. Beautifully decorated Tata trucks burn past like reincarnated rhinos. People flock without end. Colorfully clothed, beautiful people, young and old. I'm alien and pale; I stand out.

We walk past a Hindu temple and I wonder how Iceland would have developed if Christianity had not arrived around the year 1000 CE. Our gods would certainly be as diverse as those of Hinduism. There are holy icons: blue Krishna; many-armed Kali, the goddess of time, annihilation, and creation. There are Shivalinga votary symbols; there's Kamadhenu, her feet depicted as mountains, as the Himalayas; there's the elephant-headed god Ganesh, flashing disco lights all around him, plastic flowers, incense, an ancient radio.

We move on: a blind woman stretches out her hand and a funeral procession follows her, a woman resting on a bier under yellow flowers. To me, seeing this dead human is somehow disconcerting. All around us there's astounding life but also pronounced death. Dead rats, dead dogs, a man who appears to be dying curled in a ball. A woman at his side, breastfeeding. All blending together; I sense a continuum of many thousands of years. Everything new is also somehow ancient; after a while I begin to find a flow in the chaos, becoming aware that everyone is following some kind of dance or rhythm. I can't cross the street until I learn the dance. I take one step forward, then a truck driver steps on his brake and so I take my next step and a motorcycle brakes and so on and so forth. Society seems more complicated and sensitive than what I saw in China, with its rooted class divisions, religion, and traditions, none of them diluted by the Great Proletarian Cultural Revolution. But billions are waiting to rise from poverty. The swiftest way to achieve this is with our current technology. By adding coal to the fire, by erupting several more volcanoes, by increasing the flow of the river of oil.

We fly to Dharamsala on a Fokker plane. Everything is a bit more comfortable and spacious in the mountain villages; the air's fresher. By the sides of the roads there's lots of plastic and trash. Everything is somewhat shabby and yet still enchanting. Foreign travelers take care not to make eye contact and to ensure that no Westerner enters a photograph; they want to return home and claim to have experienced something unique and unknown. Everywhere there are young men in floor-length burgundy robes chanting morning mantras; some sitting in cafés with mobile phones while the holy cow munches on a trash heap. We wander around the monastery gardens. We hear an uproar—shouts and raised voices—like a group of men arguing. It turns out to be pairs of monks practicing argument techniques, almost like a martial art of logic and rhetoric. One stands and questions while the other sits and answers. The one asking questions does so in Tibetan, according to the Bodhisattva path. The other answers; if the answer is wrong, the former yells and strikes out with his arm, hitting the air and not the other monk.

We walk past the monks and approach the Dalai Lama's home, where he and the exiled government have been since fleeing Tibet in 1959. We go through a security check. Monks greet us and take us to his guest room, which is decorated with the traditional splendor of Tibetan Buddhism, in golden, blue, and red colors. Mandalas and patterns and symbols that have ancient roots in the mountains of Tibet. His assistants welcome us. The Dalai Lama walks in and sets a white scarf on me. He wanted more time to look around when he was in Iceland, he says. He wants to return.

I tell him I'm still working on a project about Audhumla, about time, the glaciers, and the holy waters.

"Ah! Yes, the magic cow," he says, and laughs.

I say I'm writing about everything that's effecting change in the world.

"You're the fourteenth reincarnation of the Dalai Lama since 1391, yet reading about your life since 1935 it's like you've lived ten lives in this one incarnation."

"That's true, to some extent. I was born in a poor village and spent the first five years of my life there, then I was brought to Lhasa and officially became the Dalai Lama, the Tibetan spiritual leader. But in reality, I was just a restless little monk. I only wanted to play, not study. Studying proved

a great burden. Therefore, my tutor had to have a whip for this restless young monk. Back then, we studied together, my older brother and I; the tutor had two whips! One ordinary whip and another, yellow one, a holy whip for use on a holy student! I was a restless, stupid student; the tutor had to use the whip pretty regularly."

He leans toward me and laughs.

"But I had a brain that knew the holy whip didn't cause me sacred pain, only normal pain. Ha ha! And so, fearing pain, I kept studying. That was one lifetime!

"Then, when I was sixteen, Chinese communists invaded and I lost my freedom. So began another life, another time, that lasted nine years."

"You engaged in diplomatic relations with Chairman Mao during this period."

"Yes, of course. I met Mao Zedong in 1954 and 1955 on several occasions and he treated me like I was his own son. Very close relations developed; I had a great admiration for him. Early on, there was a little bit of suspicion but afterward I reached Beijing and had several meetings with Mao and many others, what to call them, freedom fighters, communists, wonderful folk, dedicated to the well-being of the people, particularly the working class. I became very attracted to socialism and Marxism. I even asked the Chinese authority if I could join the Chinese Communist Party. That kind of faith and trust developed. When I returned from mainland China, I was convinced Tibet could develop with the help of Chinese communists under Chairman Mao's leadership. Then there was the uprising in 1956. I wrote to Chairman Mao. He'd promise: 'Any problem, write to me directly.' I wrote at least two or three letters and no reply. My trust grew less and less.

"I went to India in 1956 and met with Pandit Nehru. We developed very close relations and I had many discussions with him. At the time, many Tibetans suggested to me I should not return to occupied Tibet. This was a golden opportunity, a free country, better to remain there. I discussed it with Nehru and finally he advised me to return to Tibet. You have had a special relationship with the Chinese Central Government, he said. On the basis of your special agreement with them, you should carry on the struggle, within Tibet.

"Then came the Hundred Flowers Campaign in 1957, when one was meant to listen to a hundred different points of view. That movement finally brought disaster. Subsequently, intellectuals and anyone not towing the party line was persecuted. Chairman Mao often used to say, 'Communists must receive criticism from both within and from outside, otherwise we'd be like fish out of water.' That's what he said, but he did the opposite."

"And did he become like a fish out of water?"

"Trouble flared and Mao became more concerned about power rather than ideology. I usually describe myself as a Marxist, with great admiration for a Marxist economic theory, equal distribution, for not only thinking in terms of profit, the way capitalism does. Sometimes I think that the noble Marxist ideology was spoiled by Lenin's hunger for power. And Stalin, of course. And Chairman Mao. In the early period, Mao was a very good communist but later power spoiled him. That's my experience. Whether or not that's correct, historians can do the research!"

"And you fled to India in 1959?"

"For nine years, I tried to achieve peace and to cool the situation, but I failed. In 1959, some unrealistic reforms were carried out in the eastern part of Tibet, an area which was under Chinese jurisdiction, as well as some Chinese provinces, Gansu, Sichuan, Yunnan, and Qinghai—the uprising happened there and spread throughout all Tibet. Things became out of control, especially after March 10, 1959. I did my best for a whole week but I failed. I left my place in Lhasa on the seventh night and eventually reached India. So from around April 1959, until now, I've lived another, different lifetime.

"I was sad. A homeless person. But I usually tell people: Homeless, I found a new and happy home here in Dharamsala. The Indian government extended a warm welcome and gave us political asylum. Not only that: the government provided the entire Tibetan refugee community with land and, especially, education for our younger generation. Initially, all the expenses were met by the Indian government. The last fifty-one years as a refugee have, for me personally, been the happiest period. Completely free. I can go everywhere, even to Iceland. I can speak freely; no one controls me. I love that."

"But your ultimate goal must be to return home?"

"Yes. For every Tibetan."

"Have you visualized going home?"

"Yes, it's very possible. Because the problem is political. The political isn't based on realistic politics, but on a very narrow-minded perspective, shortsighted political thinking in the minds of hardline Chinese communists. As soon as they change their minds and become more realistic, the Tibetan issue will immediately be solved. No problem. In the meantime, in China, there are over two hundred million Chinese Buddhists. And many Chinese people come here, in spite of many difficulties, some quite secretly, for teaching and instruction on Buddhism."

"You're very lighthearted. How did you feel when you were sixteen and the whole responsibility for all Tibet was put on your shoulders?"

"Anxiety, of course. During those years, I had no experience, no education on temporal matters. The situation was terrible, so difficult that I built up a lot of anxiety. But I had very good, trusted advisers. I'm the type of person who can easily get along with everybody. Even in serious matters, I often asked the opinions of those who swept the hallway floors. They had open minds, they were trustworthy, and they formed thoughts for themselves. They also heard all kinds of rumors from outside and brought me news. It could be very helpful sometimes!"

He looks at me and laughs with his whole face.

"As a result, I have a kind of confidence, even when a situation is difficult, taking a decision and not worrying about it after. Even if some mistake happens, there's no regret. I had consulted thoroughly. We are Buddhist practitioners so we also have secret, spiritual methods in order to investigate. You won't understand! Ha ha! Quite mysterious! But according to my own experience, from age sixteen until now, approaching eighty, all this mysterious kind of investigation has been very accurate. I have full confidence in these methods."

I'm curious about these mysterious investigations.

"Can you give me some insight into them? About perceiving the future and doing the next right thing?"

"It's a scientific method. First, we use our intelligence, our rational ability to thoroughly analyze a situation. Then we ask people, consult. If there's some universal agreement, no need for mysterious investigation. But in the end, if you're in a dilemma about what to do, then I use this mysterious way."

"Like some kind of oracle?"

"An oracle is different. I treat the oracle more like my adviser. Asking for advice, views, but not seeking a final decision. The final decision I get from farther above, higher, using this mysterious way."

"Have you written your letter about being reincarnated?"

"In the previous or next life?"

"The next life."

"No, no. The Dalai Lama has not in recent cases sent such a letter although some earlier Lamas did, sometimes saying where they'd be born, even their parents' names. But that didn't happen with the thirteenth Dalai Lama or his predecessor, not as far as we have documented.

"Some people in the West have the impression that the institution of the Dalai Lama is so important for Tibetan Buddhism, but that's not the case. Tibetan Buddhist spirituality and culture will live as long as the Tibetan nation remains with certain freedom. In the last sixty years, including the Cultural Revolution, we've seen systematic destruction, but our culture has never been eliminated. Because of spirituality. Once it is established in the human heart, no force can eliminate it. We're now up to three or four generations who have grown up both inside Tibet and China, within China's centuries-old traditions, and we have also seen China lose centuries-old traditions. But the institution that is the Dalai Lama: it came into existence at one point and at another it may cease to exist. No problem. But if people want to keep this tradition up, they will carry that responsibility. That's not my responsibility. After my death, I'll watch from my mysterious level and see what they are doing! Ha! Ha! More than that, I have no direct link of any sort!"

"And as you face death, are you scared or curious or . . . ?"

"Sometimes curious. Not afraid. The one thing is that if I die under these circumstances, I think millions of Tibetans and many of my friends will feel really sad. That's why I sometimes feel . . ."

He thinks for a little while.

"Otherwise, personally, after all, you have to go. No one escapes from death. That's a reality. Death is part of our life. So: no problem. What's important is that so long as you are alive, your life should be meaningful, sensible, something useful to others. Then, at the end, you will have no regret."

"Millions of Tibetans await your return. Is there nothing in these mysterious questions you have been asking, any confidence that you will get to return to Tibet?"

"Yes, and not just in my mysterious investigation. Many other mysterious investigations all say that everything will finally be bright, it's only a question of time. Some predictions were made a few centuries ago but it's quite clear that this adversity is temporary. In history, sixty years is not a long time, is it? Anyway, I think this narrow-minded, authoritarian communist system has no future. That's clear."

"It is difficult to be narrow-minded for decades. Maybe in sixty years . . ."

"Sixty years have already passed. Look at the Soviet Union: it lasted about seventy years. And all the former Eastern European countries have completely changed. There were once dictatorships in Italy, Spain, and France. And China is changing, because of education, because of the Internet, all these things. No government can control things completely."

"You're hopeful."

"Today's reality is completely different from the '60s, the '70s, the '80s. In the interior of China, there are still a lot of poor people, but China has otherwise changed a lot. So many young Chinese have new fashions, copying American lifestyles. Ha ha ha. Things are changing. One important Buddhist concept: things never stand still, they are always moving and always changing. As I mentioned before, things are interrelated, interdependent. In China's case, in the '50s and '60s, even in the '70s, the leaders preferred isolation. Now that's completely changed. Previously, whatever the government said was believed. But since the days of Deng Xiaoping, a large number of students have been sent abroad and many companies from outside have come into China. Many Chinese are using both eyes and both ears. Previously, it was just one eye, one ear. Truth is always in the long run much stronger than force. That's my logic. Basic human nature is much stronger than an artificial system. Those cannot remain in place."

"So, all things considered, you think the world is improving?"

"In 1996 I had an audience with the Queen Mother of England. I'd known her from pictures since childhood and when I met her, she was already ninety-six years old. Since she'd observed a whole century, I asked her if humanity or the world were becoming better or worse, or remaining the same. Without hesitation, she said to me, better. As one example, she mentioned that in her youth, the idea of human rights and the right of self-determination were not common. Then she mentioned that nowadays these things are universal. So these are indications, she said, of the world improving. Although she did not mention it, I think it is a reality that in her youth Britain was a colonial empire. Later on, all these colonies have become independent, just like that. Democracy, has become much stronger."

"My children will hopefully still be alive in 2100. Have you tried to imagine the future?"

His Holiness leans back and says:

"I always tell people that in my conviction it is impossible to look a few thousand years ahead. Impossible to predict. But for the next few centuries, for sure, humanity will remain on the planet and at least this century, the twenty-first century, can be a century of peace. But peace does not mean an end to problems. Problems are bound to remain. But peace is about changing our attitudes to those problems and how we deal with them. In the past, whenever we met some problem, usually our first reaction was to solve things by force. That's outdated. Twentieth-century people did that. According to historians, two hundred million people were killed during that century's wars. War is not play; it has a certain purpose. One side is victorious in destroying the other side, the opponent. When war happens, violence happens.

"So the use of maximum violence, including nuclear weapons, failed to solve problems or achieve our goals. I think instead about the disappearance of the Berlin Wall and the change of authoritarian systems in Europe, not because of nuclear weapons, but by popular peaceful movements, through awareness, through an experience of suppression. Therefore, I am quite sure that the twenty-first century can be a century of peace. Peace through dialogue. In the latter part of the twentieth century, although the world was ideologically divided, two military blocks, people found a new world of coexistence despite different ideas and armies. By the end of that century, dictatorships fell in Europe. They still remain in Asia, of course."

"How do you feel about the twenty-first century?"

"I believe that this century can be a happier one. Violence is the wrong method, is outdated. You never achieve your real goal through violence. For example, toppling a tyrant in Iraq: the motivation might have been noble. But the method was wrong, and so unexpected consequences happened.

"We need to educate the younger generations. The only proper or realistic way to solve a problem is through dialogue. A willingness to listen to others' views, then try to find a mutual solution. Just before the Iraq crisis happened, millions of people went to the streets to express their opposition. These are positive signs. That's why I'm optimistic for the twenty-first century."

"Do you think it will be different from the twentieth?"

"In the twentieth century there was a lot of development, but primarily in material fields. I think we lacked inner development due to the lack of awareness of the importance of our emotions and mind. It's vital to attend to our feelings and thoughts. This century, I see more and more well-respected scientists realizing the importance of our emotions in our lives. And many people who deal with education feel the development of the brain alone is not sufficient. On my travels, I've met more and more educators and teachers who say: We lack education for the heart, we need to teach warm-heartedness. They ask: How can we introduce this new education system into our existing modern education?

"They cannot bring moral education on the basis of religious belief. Ethics and compassion must be taught on a secular basis, otherwise it will cause problems in multireligious, multicultural communities, like in India. That's why the Indian constitution itself is based on secularism, because of their reality. Secular doesn't mean disrespect for religion but rather respect for all religions. No preference for one or the other. We must also respect the nonbeliever, too. I think we can educate people and connect education to warm-heartedness. I think there are signs of hope for this century, for the future."

"It's easy to teach two plus two, but some of my teachers would have had a problem with teaching warm-heartedness. How do you teach compassion?"

"It's subjective, of course, based on each individual's experience. No instrument can measure it. But the latest scientific research into brain activity has shown that a calmer mind and positive thinking have a measurable effect on the body, while stress, hatred, and anger can weaken the immune system. This is an objective reality. And it is obvious, if we use our common sense about society, that a compassionate family enjoys greater happiness. If the father or the mother is an angry type of person, always shouting, all the members of the family suffer. That's very clear, even in animals: a dog that is always barking at others remains isolated.

"You can make a friend through power or money, but these are artificial friends. We humans are social animals; friendship is key for social animals. Friendship based on trust. Money cannot bring trust; it creates more suspicion and desire for exploitation. People cheat other people for money. Real trust comes through compassion and respect. To build trust, you need solidarity, openness, truthfulness, and honesty. All these come from warm-heartedness. Using such logic, we can teach this to other people. Warm-heartedness is so important for your own interest. The other day I heard a BBC discussion all about the power of guns, economic power, the power of truth. People love power! Some believe power comes from guns. Like Chairman Mao once stated: In the short term, it's powerful to carry a gun."

He aims his finger at me.

"So! Everyone has to listen! But in the long run, it's counterproductive. If you rely on the power of the gun, trust is destroyed. Friendship is destroyed. If you rely on gunpower, the rest of your life remains negative. Finally, it's better to eliminate weapons. The power of money might last a little longer, but that cannot bring real human friendship either. The power of compassion, the power of truth, these are the basis of our happiness. And if you just use common sense then you will see how valuable the power of compassion is.

"Your grandchildren, when they're grown up, I think they'll find the world more peaceful. I think our awareness of the importance of ecology is increasing. These are important signs. That's why I think the twenty-first century will be happier."

"But where does such compassion come from?"

"From our own experience. We're born from a mother. Although I think we're born from a cow, actually, according to your thesis!"

His Holiness points at me and laughs.

"Even if it's a cow, okay, it's a very compassionate cow! Our ultimate source, the mother, is a symbol of compassion. We survived because of a mother's compassion, her infinite love. Research shows that those who experience, in their early years, compassion and love as part of their upbringing are much happier. Those who lack from an early age an affectionate home atmosphere and grow up amid indifference and, worse, abuse tend to have difficulties and can live all their lives with a sense of insecurity."

"How do you achieve maternal love in a society of monks, where mothers can be far away?"

"You receive monkhood after reaching seven, after seven years with your mother. So there's no problem. I was, I think, about five or six years old when I separated from my mother, but she remained quite close. My mother came every day with some special village bread. Oh, my mother was expert at making that bread. So, you see, we were very close. Even after we fled to India, when I was about thirty. I always received my mother's affection.

"When I talk about secular morality, I use three criteria: common sense, shared experience, and the latest scientific research.

"In all religions, there's a shared, compassionate thread. The importance of love, forgiveness, and forbearance, despite different emphases or attitudes. That's how compassion is borne in our blood; now the time has come to tend to it, to cultivate not only material things and the intellect, but to nurture the heart. When warm-heartedness merges with wisdom, society will become stronger, more fortunate and more affectionate. Problems will decrease; when they do appear, we can solve them collectively. Not with hatred and suspicion and by separating people or threatening one other. That only creates more character problems. It's very sad."

"Should we extend our compassion toward nature? In Buddhism you try not to harm life but suddenly in this interconnected world of ours we never know what organisms we are actually harming."

"We cannot have a compassionate attitude toward a non-sentient being—like trees or grass or plants—that has not developed a sense of compassion. But respect is closely related. Every living thing has the right to exist or to survive. All these plants are a part of nature, broadly speaking, and we are also a part of them. We can't survive without them. And then there are birds and animals that, from a Buddhist viewpoint, are living, sentient beings, so we must extend our compassion, our love toward them."

"The world today is entirely interconnected. Recently there was a volcanic eruption in Iceland that had a serious impact on floriculture in South Africa. Did this affect you?"

"No, fortunately I wasn't traveling. But if I'd been visiting many countries, I would have suffered! And maybe would have complained to Iceland! Ha ha! Bad Iceland!"

"For us, it was almost like suddenly we had some power for a moment. The world stopped because of us."

"But power you could not control. That's a pity!"

"True, but we were almost a superpower for a few weeks!"

"That reminds us that no matter how sophisticated our technology is, ultimately we are subject to nature. That's an important reminder. Global warming can get beyond our control. We must take as much care as possible. The future of seven billion human beings depends on nature. It's very important to realize and accept that. Sometimes, very sophisticated, advanced technologies and science give us a false confidence that we have the ability to control nature.

To some extent we can but beyond that, we have to live in harmony with nature."

"What about Tibetan independence?"

"The Tibetan issue is essentially a man-made issue, essentially created by our eastern neighbors. So we have to solve the issue with them. Like other human problems, as I mentioned, we have to find ways and means of solving this issue through dialogue. As for Tibet, so for Palestine. You cannot solve such problems by taking a tough stand, where one side wins and the other loses. That method has never achieved any lasting solution.

"The Tibetan way of thinking, through Tibetan Buddhist culture, is more peaceful. In the long run it can be very helpful to bring about a more peaceful society in mainland China. Considering various facts, we believe in the middle way, not seeking independence but being part of the People's Republic with certain constitutional rights and some documents on the rights of minorities. If Chinese communists want unity and a harmonious society, many of their policies are actually unrealistic. They have the wrong method, trying to achieve unity by force. How can they? Impossible! You might be able to bring together hundreds of cows with one whip, but we're human beings. Real unity and harmony come from the heart! Fortunately, among Chinese leaders now there is more open-minded thinking. And particularly Chinese writers, intellectuals, professors who feel their present policy is wrong. These voices are emerging. That's a hopeful sign.

"Setting aside history, our stand is: the past is the past. No matter what past history is, we are looking forward. Like the European Union: they established the European Union not on the basis of past history but according to the new reality, they developed the idea according to a long-term future. Similarly, we're not thinking about past history. Simply about the future."

"Do you think you will be able to get back to your home country soon?"

"Yes, we believe so. Particularly in the last sixty years, you see the change in Chinese communists. I usually say there have been four eras. Mao's era, Deng Xiaoping's era, Jiang Zemin's era, Hu Jintao's era. Those four eras have big differences. They show that the same Communist Party, the same one-party rule, has the ability to act according to new realities. So I'm hopeful.

"Anything else you would like to ask?"

"There are two monks out there debating in the square; the sentence they're discussing is this: If my desires cannot be fulfilled, even by everything here upon this Earth, what else will be able to satisfy them?"

"One important thing I am always telling people: materiality is insatiable; there's no contentment, just more, more, more. Mental development, mental health, nourishing yourself: that you can develop indefinitely."

"Thanks so much for the conversation."

"Thanks. See you again. Maybe in Iceland. I really want to go there once more."

In a mother's milk

At the Tibetan border, in Nepal's Mustang province, the Nhubine Himal glacier is almost 6,300 meters above sea level. Beneath it runs a milk-white river called Kali Gandaki. The river is so ancient it is older than the Himalayas themselves, and they formed fifty million years ago, when the Indian tectonic plate drove slowly and steadily against the Asian plate. While the mountains lifted up to the heavens, the river buried itself so that it now flows through the deepest gorges in the world, where the bottom of the river is 2,500 meters above sea level yet the highest peaks on either side reach over 8,000 meters high. On its long journey through the Himalayas, the river accumulates streams and branches and other rivers that have their source in more than a thousand glaciers.

I reach the river six hundred kilometers from its source. By that point, it's become a broad, calm, mature river, known as the Gandak or Narayani River, flowing through sandbanks and gravel, through the Chitwan National Park in Nepal. There, the river's name is Narayani; farther on it joins another river, by which time the water has become fully sanctified and is known as Ganges, India's mother herself. Here, for the first time, I encounter the holy water that springs from the frozen teats of Audhumla in the Heaven Mountains. I'd thought about going up to Kailas and walking around the mountain, but it's probably best to respect its holiness. Trying to see and experience everything is a bad habit. Some things should just be left alone.

Crossing the river to get to Chitwan National Park, I saw a gharial crocodile basking on the sandy riverbank. I was strangely pleased to see this old dragon with its peculiarly narrow snout, specially designed to catch fish in the white glacial water. No more than two hundred animals of this species remain alive in the wild; their global range has diminished to just 2 percent of its previous extent, when gharial crocodiles flat-bellied in their thousands in all the major Asian rivers. I also saw a mugger crocodile, more traditional in appearance, slightly more dangerous, and found in small groups in Iran, Pakistan, India, and Sri Lanka. It has been eliminated in Burma and Bhutan. I saw odd footprints on the riverbanks. The guide said we had to take care, these were the tracks of the Asian rhino, *Rhinoceros unicornis*. The unicorn rhino.

We saw these beasts shortly after, in the reeds, a pair with a calf, huge creatures with such ancient characteristics. A male can weigh up to two tons and be two meters tall, not exactly a great animal to meet on a trail. Their skin is thick, in strange folds, the way I imagined dinosaurs when I was younger, stegosaurus or triceratops. But the rhino is much younger than the dinosaurs, no more than ten million years old. There are currently only 2,500 animals left, in tracts of land in Nepal and India that measure about twice the size of Vatnajökull. A herd of elephants bathed in the river and we followed them into the jungle where tigers could be found. I didn't see them but I spotted footprints, the size of my own footprint, size eleven, a sign that this was no domestic cat. Nepal has fewer than two hundred Bengal tigers left. I was not sure whether I should mourn the animals that remain from the orchard our Earth once was, or rejoice because the rhinos and tigers in these areas have recovered in recent years. Our tour guide, Champo, told us to be careful, although attacks were uncommon. At that moment, a silent gasp came from my traveling companion and she pointed ahead. The path we were walking toward was carpeted with four-leaf clovers. I ran my hand over the clover leaves and three little leeches clung to the back of my hand. Tiger tracks, unicorn rhinos, four-leaf clovers, and leeches beside a sacred river. It would have been interesting to interpret this dream, were it a dream.

The white sand on the bank of the Narayani River was fine as ash and, in fact, it was partly ash, from dead people: in the distance we saw a fire where a family was saying goodbye to their relative. The sun was setting, fire-red in the mist, birds were flying low over the mirrored river, and night birds were awakening. It was obvious why this water had become holy. And there, too, we could see how everything was connected. A holy, white glacial river running from Tibet's Heaven Mountains, the source of life for humans, crocodiles, and unicorn rhinoceroses.

After the interview with the Dalai Lama we went to Amritsar by car and from there by train to Delhi. We had a night out in the city; a guy knew a guy who knew a guy who invited us to a party. A black Range Rover picked us up and once we'd started driving, we heard a voice from the rear.

"Go right. Head left!"

The driver was in the trunk giving directions. The car's owner was at the wheel but he'd never driven himself around Delhi. We drove down dark, ill-lit streets where people were wandering about and where cows lay by the side of the road; we drove past a poor neighborhood where families seemed to live on traffic islands and under bridges. A group of children ran across the street in the middle of the night. It was as if Dickens's *Oliver Twist* was a contemporary story. We drove through some massive gates and along a driveway where hundreds of luxury cars were parked, and then we walked through a colonnade into a brand-new house that looked like something out of a Disney version of *The Thousand and One Nights*. We went into the lobby; off to the right there was a sumptuous room with huge paintings on the walls and crystal chandeliers hanging from the ceiling; three men were sitting at a table playing poker. "I'm good at poker," our friend said; she asked if she could join the game. The bid to enter was $100,000.

We left and walked toward the music to find the party out in the garden. There, the young Delhi elite amused themselves, a DJ playing thundering techno while eighteen chefs and waiters served delicious food. Our host welcomed us. He was quite a cool guy; he'd lived for a time in London and Dubai but now lived in New Delhi. He was wearing a T-shirt with the inscription *Will Fuck for Coke*. There was a large swimming pool and at some point the bass kicked in and everyone threw themselves fully clothed into the pool.

On the other side of the garden wall was a one of Delhi's biggest slums, where a tanker brings water once a week and people get into fistfights with their bottles and pots. It was a tangible manifestation of Mahatma Gandhi's words: "Earth provides enough to satisfy every man's needs, but not every man's greed."

The inequality was obvious because only a simple wall separated these worlds. Absolute destitution on one side, complete excess on the other. This made it easy to point at the situation, to condemn. But what does it matter if poverty was on the other side of the wall, the city, the mountain range, the border, the high seas? I knew which side of the wall I was on, at least.

When Grandpa Árni became a poor, fatherless child in Iceland, the prevailing ideology brought the destitute widow to the most modern apartment in town; that quite possibly saved his life. Unfortunately, we still live in a world where many such children are left out on the street.

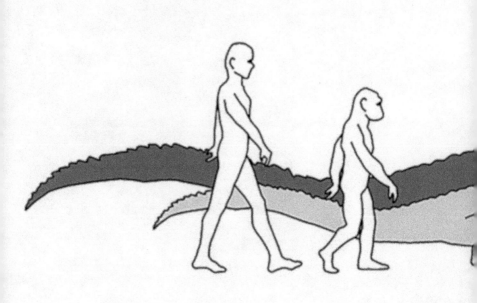

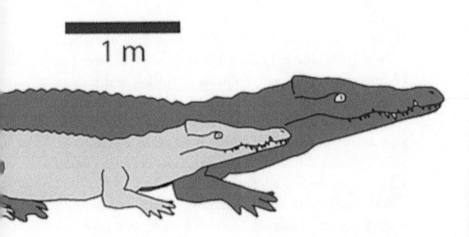

Crocodylus niloticus　　　*Crocodylus thorbjarnarsoni*

Crocodylus thorbjarnarsoni

My maternal uncle, John Thorbjarnarson, died of malaria in New Delhi, India, in 2010; his ashes were scattered in Florida's marshland amid alligators, turtles, and roseate spoonbills. It occurred to me that if Earth were at all sentient, willing, or had an immune system, she would spare people like him. Advocates of wild nature, of swamps, of little-loved animal species are, sadly, far too few. He was only fifty-two years old and he had a lot of work left to do.

Although John died, his work in thirty countries ensured that his passion and expertise thrive via a large crowd of scientists working to nurture and research fragile crocodile populations. I've read scientific reports and scholarly articles referring to John's guidance in Uganda, China, Brazil, Venezuela, India, and elsewhere. According to the *Economist*, his nickname in Burma was Kyunpatgyi, after a mythical giant crocodile that according to legend swam around an island in the Meinmalha Kyun Wildlife Sanctuary. By protecting crocodiles, humans preserve wetlands and habitats for other species. Wetlands also form natural flood protection and store carbon. Everything is connected: wildlife habitats directly affect our future prospects. Biodiversity is the axle hole that determines the utility of the wheel. As the *Daodejing* says:

> For existence to bear fruit
> what does not exist is most useful.

At Lake Turkana in Kenya, scientists have been working for years studying sediments that provide information about the dawn of humanity, the origin of *homo sapiens*. In the sedimentary layers, they've found remains of many species that existed five million years ago, along with the first humanoids. In 2012, scientists from Cincinnati researched the skulls of a previously unknown crocodile species found in sediments from the Pleistocene era.[76] Through further study, it was discovered from the size and shape of the skull that this was probably the largest crocodile species known to have ever existed. The adult animal was almost eight meters long, making it the largest predator in the region. This crocodile was a true dragon; our earliest ancestors were probably part of his prey. To honor John Thorbjarnarson's memory, this new species was named *Crocodylus thorbjarnarsoni*. You might say my uncle has been reincarnated as a prehistoric crocodile. I reckon this pretty much fulfills his childhood dreams.

I also think John showed me it's possible to nudge the world. That the world is not just an out-of-control and meaningless flood, always in flux; it can be influenced, can be steered in the right direction. Our purpose is to be useful, to make a difference, to increase knowledge, to point the world in the right direction if it's off course. John's collaborators in Uganda still keep his memory alive; research continues into the habitats of the dwarf Nile crocodile. Carol Bogezi is one of those who works in his wake and acknowledges his influence, a young woman who is a pioneer in animal welfare and women's rights in Uganda. Not only did she have to fight prejudice against crocodiles, but also against women: her research involved having to lead a team of twenty men not accustomed to taking orders from a young woman. Bogezi has worked as a mediator between wild animals and humans, and according to her latest article, she is beginning to expand her knowledge and explore how farmers in the state of Washington in the U.S. can come to tolerate and live alongside wolves. In an interview, she was asked when she got the call to care for wild animals; she traces her dream back to her childhood.

John, who had crocodiles on his brain as a child, is a fine example of how early some people find their path in life. As a child, he saw a documentary on endangered crocodiles and ten years later he began working on their protection. Twenty years later, he'd played a major role in saving an entire animal species. I'd warrant his work comes close to the Dalai Lama's definition of a life that has served a purpose.

2050

*Humans are now starting to take revenge. Fire,
water's main enemy, allows them to drive their ships
against storms and currents, between breakers and
reefs; who knows how long it will be before they rise
above the waves, and start sailing the skies?*
—Fjölnir, issue 1 (1835)

If someone a hundred years ago had looked ahead and
assigned himself the task of orchestrating the next hundred
years, it would have seemed absolutely impossible. To provide for seven billion inhabitants on Earth, to found the
United Nations, to feed and clothe people, to educate and
house them, to connect the world via telephones and computers. To find energy sources, vehicles, and employment
for the entirety of this population. To establish thousands
of symphony orchestras and to cure previously incurable
diseases. We are a long way off from ensuring that everyone
fares equally well, but entire continents have risen out of
abject poverty in a surprisingly short time.

As the twentieth century began, humankind had not
apprehended the principles of aeronautics. In 1903, the
Wright brothers flew an airplane thirty meters, the longest
distance an engine-driven vehicle had managed to glide.

Two years later, they'd mastered the technology and were able to fly more than thirty kilometers. This achievement stampeded around the globe so that by 1917 the Red Baron was waging aerial battles in biplanes and triplanes all over Europe. In 1927, Charles Lindbergh was the first person to fly across the Atlantic, traveling from New York to Paris. There are currently about ten thousand airplanes simultaneously in the skies throughout the world, with about a million passengers on board. At the time my grandma was born, no one had crossed the Atlantic on a plane; forty years later, people had reached the moon.

Nuclear energy developed even faster. From the time British physicist James Chadwick theorized the existence of neutrons in 1932, it took only two years for Enrico Fermi to use neutrons to split an atom. Five years later, scientists conjectured that a nuclear reactor could start a chain reaction, which they demonstrated under the bleachers of a sports field in Chicago in 1942. At that point, nuclear bombs were but a theoretical possibility. The Manhattan Project was immediately launched and ten thousand people were sent out to the New Mexico desert under Robert Oppenheimer's stewardship in order to fashion a bomb. It was completed in late July 1945 and unfortunately used three weeks later. That isn't so long ago. I know a person who operated on Oppenheimer and I've talked to an elderly Japanese man who suffered the bomb at Hiroshima. The time from the scholarly hypothesis of the existence of neutrons to production of a complete bomb was less than thirteen years. This horrible fact nevertheless demonstrates what people can achieve when they think life and death are at stake.

The last thirty years have been the age of computers, telephones, the internet, mass entertainment. If in 1990 I'd wanted to buy everything that now comes as part of my phone—location tracker, computer, modem, calculator, video player, music player, navigation system, movie camera, library, conferencing equipment, phone book, phone, fax machine, game console—it would have weighed tons and cost a fortune. The speed at which processors compute has doubled every two years and that processing speed has altered our brains. These rapid advances are magnificent but they bewilder us. We wait for big problems to be solved by a genius in Silicon Valley, preferably by providing us with a user-friendly app for our phones. Magicians have hypnotized us with shiny gadgets and ever flatter screens. The world has become connected, multiplying the flow of information and access to entertainment, but our devices are designed to cultivate addiction, to immerse us in a vicious cycle of surveillance and consumerism. We are farther from nature than ever before, if nature can be measured in the amount of time a child spends playing out in the open air.

While virtual reality has waxed, reality has waned. Western powers have reduced their emissions but more often than not manufacturing and pollution have simply shifted elsewhere. Consumption has increased exponentially, generating massive amounts of garbage, the destruction of rainforests, food wastage, fashion waste. Humans own more cars than ever before, buy more stuff and use it for a shorter time, trash more food and clothing than ever before, use more plastic and fly more often for trivial reasons. The biggest developments of the last thirty years are setbacks when we consider the atmosphere and Earth itself.

The world's children have begun a climate strike and, as if in some old mythic story, the child Greta appears, fated to tell us nothing but the truth. She's related to Svante Arrhennius, the man who first calculated, back in the late nineteenth century, how increased carbon dioxide in the atmosphere might cause global warming. He actually believed such warming would be beneficial and that it would take us hundreds of years to increase CO_2 volume by 50 percent. He underestimated how vigorously we were going to burn coal and oil. It took us less than a hundred years, but his main conclusion was still correct. In the summer of 2019, heat records were set all across the world, forest fires blazed in Siberia and Australia, droughts threatened the peoples of Africa, India, and elsewhere. Previously, people feared that the world's oil wells would run dry; now studies show that if we burn all our oil, the world will burn, too. The more CO_2 we produce, the greater the likelihood of reaching a tipping point, a moment when unrestrainable processes will have begun, beyond anything humanity can deal with.

Today's children demand their education take into account the challenges facing humankind, demand that the world's nations respond to the scientific community's urgent warnings. American writer Neil Postman discusses the crisis of the education system in his book *The End of Education*. There, he argues that the education system has always served a higher purpose, what he terms "gods." First, it served God Himself in the monasteries; then kings took up the baton and the system served them; then republics and nations became the new gods. In these late-capitalist years, the education system has focused on the individual, creating human resources for business, part of an international competition between large corporations in the free market. Profit and growth are at the center of things, but do not yield to a higher purpose. Why learn? To get a well-paid job. To create more growth, to pour more oil on the flames faster, to go full speed ahead right over the edge.

A new paradigm has come to center stage: the future of our Earth and its atmosphere. Education systems now need to prepare a whole generation for a working life based on humans being able to coexist in balance with the very foundations of life. Why should we learn ethics? Because the coming years will be full of moral challenges. Why learn algebra? We will need to absorb hundreds of gigatons of CO_2 and no one knows how to go about that right now. Why study poetry and ancient songs? Because poetry is the silver thread of the human spirit; without it, human existence is unthinkable.

It almost requires redesigning the legacy of the twentieth century in its entirety. We must rethink what we eat, rethink trends, fads, technology, transportation, the whole of manufacturing and consumerism. At the same time, the Earth will have to feed nine billion people and humanity will need to preserve what remains of unspoiled nature. The world must be reenvisioned and that has to happen as fast or faster than humans developed flight, nuclear energy, and computer technology.

To counteract climate change on Earth, all CO_2 emissions must end by 2050. In the next thirty years, consumer habits need to transform; there must be a complete revolution in energy production and in transportation. Scientists believe that the CO_2 content in the atmosphere ought not to exceed 350 ppm. It is now at 415 ppm and increasing by 2–3 ppm annually. So even once emissions stop, 1,000–2,000 gigatons of CO_2 already in the atmosphere will need to be reabsorbed.[77] To put that in context: this figure is the amount of CO_2 all human activity produces over a thirty-year period.

Earth's inhabitants face a challenge only previously encountered in science fiction: a terraforming project to seize and control the proportion of gases in the atmosphere. This goal must be achieved by the time children who are currently approaching the end of elementary school get to their late forties—by the time my generation has reached retirement. The task is to save the earth; it cannot be escaped.

Because of climate chaos, an entire generation is being asked not what they want to become but what they *need* to become. That situation is actually not entirely negative: a whole generation will feel it has a role, a higher purpose. Those who want to try to "find themselves" might need to postpone their soul-searching for thirty years while the world is saved.

A large part of the solution lies buried in the imagination. It is unlikely that the new, carbon neutral world will mirror our gas-guzzling past. If metropolitan traffic jams turn into long lines of electric cars, the problem will not have been solved. Electric cars need tons of steel, aluminum, lithium for batteries. In Iceland, traffic emits a million tons of CO_2 a year. How fast could that change? If everyone holds off on using their car every tenth day, we can achieve a 10 percent reduction in CO_2 emissions, immediately, tomorrow—15 percent if people don't use their car one day a week. Could we imagine cities in thirty years' time utilizing other transportation systems than the ones we're familiar with today? Public transport and light vehicles—scooters, electric bikes, ultralight cars—could replace a huge portion of today's automobile fleet. We don't have to crawl into our shells every morning as though we're hermit crabs. With electric power and public transport, we can reduce traffic emissions to zero. Individual awareness alone will not be enough, however; Reykjavík's geothermal heating system didn't come into being because private citizens constructed their own geothermal wells. The largest and most vital solutions are only achievable through a combination of government initiatives, the contributions of visionary politicians, and international cooperation.

Scientists around the world are looking for solutions, many of which will be unexpected. A team of experts who belong to a group called Project Drawdown have compiled a hundred key climate solutions and ranked them by gigatons.[78] The UN's Sustainable Development Goals similarly provide a guide as to where the world might go. Most studies orient us in the same direction and the solutions can be roughly divided into four categories:

1. Reducing food waste and make dietary changes.
2. Developing solar and wind energy; electric-powered transport.
3. Working on conservation of forests; afforestation; restoration of wetlands and rainforests.
4. Empowering women.

What's hopeful about all these solutions is that they represent a unified vision for a better world.

Changes in our climate are starting to affect harvests all around the world; the Earth's current fertility is no guarantee of its future. Agriculture is unsustainable in many countries and soil has deteriorated for some time due to the adoption of harmful methods of cultivation. We waste about 30 percent of the food we produce while a large part of the world's food production goes toward feeding livestock. Studies show that the United States could produce food for eight hundred million people if all the grain that ends up as animal feed were used for human consumption.[79] Humankind could sustain about four billion more people using just currently available agricultural land, without needing to encroach on the remaining rainforest or Earth's unspoiled spaces.[80]

One meal of beef is equal to twenty meals of pasta, based on carbon footprint. Perhaps Hindus were right to accord cows sacred status; if cows were holy worldwide, our problems would be far less serious.

In many cases, we do not need any new technology to solve our problems. The key to the solution is protecting nature. The industrialization of the last centuries has reduced rainforests and wetlands to useless land or to mere raw materials; nature reserves have been established so people can sample a world that once existed. Rainforests and wildernesses play key roles in capturing carbon and protecting the atmosphere. Their existence has become essential to the lives of all Earth's inhabitants.

Wetlands are vital because they capture and store carbon. In Iceland, wetlands were actively dried out on an immense scale; the ditches involved were, in total, equivalent to the perimeter of the Earth, nearly thirty-three thousand kilometers. Emissions from the drained wetlands in Iceland exceed that from all its heavy industry, automobiles, and its aviation industry, combined. About eight million tons annually.[81] In comparison, automobiles release one million tons of CO_2. As the land dries up, rot sets in, oxidizing hydrocarbons that have accumulated in the soil for thousands of years. What one generation found useless was actually useful; by "improving" the farms the land became a destructive force. Halldór Laxness wrote about the excessive drainage of moorland in his 1970 article "The War Against the Land," concluding with these words: "Shouldn't we make a case for paying people to fill them in again?"

About 70 percent of the drainage ditches in Iceland aren't used to help cultivate hayfields. Researchers have found that Halldór's proposal to fill in these ditches and restore the wetlands would be Iceland's most important contribution to preventing climate change.

In recent years, significant progress has been made in solar and wind energy. New wind farms and solar fields have become competitive with the old coal-fired power plants, and rapid advances in battery technology are being used to offset energy fluctuations. If U.S. electricity needs were met with solar energy, it would require about ten thousand square kilometers of space, about the size of the Vatnajökull glacier. Roof space in America measures about five thousand square kilometers while parking lots cover about sixty thousand square kilometers.

In this race, everyone wins, or everyone loses. The sooner Australia, Arabia, and Arizona are powered by solar energy, the more likely the technology will be widely available, accessible to people rising from poverty. In Africa's remotest areas, there are people with smartphones who skipped a hundred years of advances in communication technology.

The same thing must happen in energy innovation. Skipping coal and oil and going directly to solar, wind, and thermal energy. It is a realistic development if richer countries scale up the technology and utilize it themselves.

Some might think that women's empowerment is not an environmental issue, but research shows that girls' education ensures families' well-being and that the best way to cope with population growth is to allow women to make their own decisions about when and whether to have children. Thus, equality is one of the most important solutions to future environmental problems. It is not enough to know what the solutions are; we need to show measurable changes almost immediately if we are to cease emissions by 2050.

The twentieth century established much of the infrastructure we take for granted and no longer notice: electricity lines, water supply systems, sewage, heating, telephone networks, road systems. The year Grandpa Björn was born, 1921, construction started on the first hydroelectric power plant serving Reykjavík, on the Ellidaá river. Electricity has been with us such a short time. During his lifetime, available energy in Iceland has increased from 1 megawatt to 2,700 megawatts. At the beginning of the twentieth century, people arrived in Iceland with new and unfamiliar job titles: engineer, mechanic, radio operator, switchboard operator. These people pioneered their era's technology. There are three people in the Icelandic telephone directory today who list their profession as "carbon busters," a job that will probably get a more formal name, like "carbon capture engineer." That industry needs to become gargantuan. The greatest leap forward in the twenty-first century must be the capture and disposal of CO_2, or the development of methods to extract

CO_2 directly from the atmosphere and make something of value from it.

In Hellisheidi, the first carbon capture engineers in history have taken the first steps, just like the Wright brothers did with aviation in their day. The Hellisheidi Geothermal Power Plant generates about twenty thousand tons of CO_2 annually. In 2012, research was done on how CO_2 could be absorbed into the bedrock. The process involves mixing CO_2 with water, converting it more or less to seltzer, pumping it into the soil, and allowing it to come into contact with basalt. The resulting reaction forms Iceland spar, $CaCO_3$, crystallized calcium carbonate, the same material coral uses to build its shell. Initially, people did not know if the reaction would take place over several years or even over millennia. It turned out that the rock changed in just a few months and the core tests showed that the air had transformed into glittering rock. In 2014, 2,400 tons were pumped into the Earth; by 2017, that figure had reached 10,000. This method can be used almost everywhere there is basalt bedrock, including the seabed. Similar methods may be used to make building materials, concrete that can capture CO_2 and not release emissions. Once that happens, humans will have learned to build a shelter from the same material as coral or hermit crabs.

On-site at Hellisheidi, there's a small shed where a method of extracting carbon dioxide directly from the atmosphere is in development. This method is expensive at $300 a ton but the cost is coming down every year, as the project scales up. In 2017, they managed to capture and remove about fifty tons directly from the atmosphere. That balances out one three-hour flight or the annual carbon emissions of three Icelanders. The current goal is to capture 100,000 tons. But in order to combat global warming, this industry, combined with reforestation, needs to extract thousands of times more volume than that. At present, we have no good way to assess which solution will dominate, which will achieve most success. After thirty years, this paragraph will hopefully be obsolete, like a nineteenth-century author explaining the phenomena of a "water closet" and "sewage."

I met up with Sandra Snæbjörnsdóttir, one of the world's first carbon capture engineers, and our conversation soon turned to the future. She believes that within a hundred years, international conferences will be held where experts will discuss what base level we should decide on for atmospheric CO_2 concentrations: whether it should be 350 ppm or 250 ppm.

The solutions are manifold; some are absolutely beautiful. In Iceland, I talked to a man who had restored the wetlands on his farm and he described how fast nature came alive: "I filled in the ditches and made a pond. The following summer a red-throated loon appeared and I've seen fifty species of birds where there used to be just ditches and unused fields." Many solutions improve the welfare of both humans and animals; they'll lead to better transportation, improved living conditions, and better ecology. They'll lead to action, awareness, and a sense of community. The solutions partly involve what our grandmas always told us: eat everything on our plates, hand down siblings' clothes, darn socks, practice frugality. A large part of the solution requires sacrifice, doing something for others without asking for anything in return. Creating groups and collectives and organizations, in the spirit of the Icelandic rescue teams and glacial research societies. Lessons can be learned from the past, finding the point where happiness thrived before overconsumption prevailed. Somewhere we will find balance and contentment. In Grandpa Árni's case, that was probably the Heaven mountain lodge right above the Valley of Dreams.

I believe Audhumla revealed herself to me and made me write this book so as to secure my children's lives and future. I think it likely that CO_2 is a test, a trial humanity must pass, that global warming is a warning to humankind: they who destroy the Earth's wildlife, raze the rainforests, and violate the all-encompassing silence of God's great expanse will themselves be lost. Since humankind faces a shared challenge, the world's nations will be forced to work together in an unprecedented way. There is no certainty this will succeed; all things must one day come to an end, and that applies to humans like everything else. If we succeed, the world will still be far from perfect, but it will be more beautiful than words can ever describe.

A future conversation

We're in the kitchen in Hladbær. The Ellidaá meanders through a fall-colored forest; wisps of steam rise off the swimming pool at the foot of Selás hill. A raven sits on a lamppost in front of the preschool. Hulda Filippía, my daughter, bought this house and renovated it; she's now just over ninety. Her grandchildren, twelve-year-old twins, are sitting in the kitchen with her. They're munching pancakes and watching the frame on the table that plays videos. Well-mannered children are sitting in the dining room at number 3 Selás as a young woman enters carrying a cake with lighted candles.

"I knew that woman," says Hulda Filippía, "I'm named after her. She was born in 1924, 178 years ago. Let's do some math," she says, stirring pancake batter.

"What math?" the two girls ask.

"A little puzzle my dad taught me when I was ten. When is someone you will love still going to be alive?"

"What do you mean?"

"You're twelve years old. When will you turn ninety?"

They jot down on a piece of paper: 2090 plus 90 is 2180.

"Now let's imagine your ten-year-old grandchild, born in 2170: when will that person turn ninety? When would they still be talking about you?"

They work out the sums.

"Would it be 2260?"

"Yes, can you imagine that? The person you'll love most in all the world will still be alive in 2260! Imagine your time. I was born in 2008, and you'll know a person who'll still be alive in 2260. That's the length of time you connect, more than 250 years. The time you can touch with your own hands. Your time is the time of the people you know and love, the time that molds you. And your time is also the time of the people you will know and love. The time that you will shape. Everything you do matters. You create the future every single day."

Apausalypse Now

COVID-19 POSTSCRIPT

May 10, 2020

We have been stopped. I never imagined it could happen.

Once I did an experiment in Reykjavík. I convinced the mayor to turn off all the city lights for half an hour while an old astronomer talked about the stars on national radio. The idea was to bring back the starry night to children who might never have seen a deep black sky. When the lights switched off, many things were revealed. When the lights went out, the sound of the city also dampened. People began whispering, neighbours met in the soft autumn darkness and gazed into the sky. It took six years to convince the city to do this for just half an hour. Just to turn off the lights to see the stars for half an hour.

We have been saying for many years that we as a human race are going too fast. We are cutting too close to the earth's boundaries, diminishing biodiversity. In the next eighty years, we expect the ocean's pH to change more than it has in the last fifty million years. Ancient glaciers and permafrost that have been intact for thousands of years are predicted to melt in the next eighty years, as well. We have to slow down to avoid a total catastrophe. In my book *On Time and Water*, I asked this question: if we are sensible creatures and we know where we are heading, we stop, don't we? But never in my dreams would I have expected the world to be stopped so fast and in such an extreme way.

Like most people during the COVID-19 global lockdown, I experienced lost opportunities. All my lectures and performances were canceled. I had just finished a documentary film, *The Hero's Journey to the Third Pole—A Bipolar Musical Documentary with Elephants*. After three years of work, it was to premiere in cinemas across Iceland, smack on the day that mass gatherings were forbidden, and all theaters closed down. With no certain opening date on the horizon, our months of promotion went down the drain, all our interviews had been in vain, and a mental health awareness campaign had to wait while the world discussed more urgent issues of physical health. For a week we felt frozen, with our work halted, then my co-director, Anní Ólafsdóttir, and I began to wonder. Are we missing something? Are we sitting in the old world, grieving a lost opportunity, while a historical moment slips away?

The film industry was on hold, all equipment shelved, all the talent sitting at home and waiting. We borrowed a super-good ARRI Alexa camera and asked a cameraman to join us. And then we went on a journey to capture the void. Our quest was not only to capture the empty spaces but to catch the thoughts of artists and thinkers in the midst of the global uncertainty. We asked, "What's in the air?" We wanted to capture this particular moment of having no idea of where the world is heading. Is the world changing for the better or for worse? What is the meaning of this great pause—what is it showing us?

Our first interview was with Sigrídur Thorgeirsdóttir, a professor of philosophy. She spoke about *The Decameron*, the sequence of stories by Giovanni Boccaccio, which take place in a villa outside Florence in 1348 while the Black Death ran through Florence. The story is about ten young people taking shelter from plague in the villa, telling stories for ten days. Sigrídur felt that we need stories to understand what is happening.

We did not think too much about it. But, after days of intense filming, we loaded everything into the timeline of our computer to see what we had, only to find out that we had by accident captured ten people telling us stories during ten days of filming. We had filmed some kind of a modern *Decameron*, and while we were at it, the days had passed in a flash, and in Iceland the peak had flattened, and the city was opening again. We had captured a moment which, hopefully, will never be caught again.

We wanted to explore to what extent people could keep creating in this restricted era. We kept all the rules of distancing: we traveled in separate cars, interviewed through windows or half-open doors, while each artist did some kind of a performance with us or for us. The idea was that art is happening in every situation; art always finds a way, always, even when everything stops; art is possible and is perhaps even more important. The limitations shape the work; they are part of the art but not something that prevents it. By making a documentary in the void—the in-between space that was the lockdown—we kept sane and active, asking artists to do the same for and with us. But this was not art just for art's sake; it felt vital to understand what happened and our desire to explore the possibilities of this moment. What is the meaning of experiencing the big pause, the great stop, the *apausalypse*? What does it tell us about ourselves, our bodies, our nature and our systems?

Haraldur Jónsson, the visual artist, told us that the word "apocalypse" in Greek means "to uncover something." And this was an apocalypse: an uncovering of our smog and smoke, an uncovering of our fragility, our supply chains, uncovering competence or incompetence of governments, revealing to us how health is not an individual issue, because the health of every person on the planet is connected—and is again connected to the health of the earth systems.

We met Ragnar Axelsson, a renowned photographer. He had just been driving around Iceland, capturing the void on camera in the total absence of tourism or domestic travelers. The animals, he said, were different. It was like they had forgotten all about us: the birds did not move; horses were standing in the middle of the road. He told us a story from Greenland about following hunters out onto the ice. They told him, "You must never disrespect nature with words or deeds. It will strike you back."

Dancer Unnur Elísabet was supposed to finish her degree in acting in Barcelona. Taking the last flight to Iceland after the school closed, she had to cancel her big final performance and finish school with a simple online task while quarantined alone for two weeks in her parents' holiday home.

We wanted to film the empty Keflavik International Airport—our main connection to the outside world, normally crammed with people, now totally empty. We wondered if we could do something extra here, and ask the dancer to dance her way through the empty airport. We made a few phonecalls with the authorities and it turned out to be no problem. They couldn't have been more helpful. Well, why not? The airport was empty, with only two cargo flights taking off in the morning. So we asked again: "If there are no flights, can we dance on the runway? Possibly the only moment in history that will be possible." "Can I call you back?" They called back with the answer: "Well, yes. You just need our staff to escort you." So we started planning the dance on the runway and had another thought. It would be nice if we could drone the performance.

We called them again and this time you could hear a child singing in the background, as a homebound civil servant responded. "Can we drone the dance?" I asked. "Over the airport?" "Yes. Just fill in this form," she said. So Unnur danced for us. She danced away her frustration over lost opportunities. She danced away her two weeks of qurarantine, danced through Customs and security, through the duty free store out into the runway, where the only flight to Europe that day, a cargo flight, took off.

Gunnar Kvaran played Bach on his cello for us; he had been mentoring students on Skype in the last weeks. He said these events could not be a coincidence; the virus must be a warning sign from the universe. Scientists had told us what is happening in terms of the climate and what will happen if we do not take action. Governments and businesses had told us that nothing could be stopped. Now we are experiencing this great pause, and during this moment we must rethink everything. We stopped to save the lives of our parents and grandmothers, and now the question is: can we make a similar effort for the climate in the name of our grandchildren? He said that now, in his late-seventies, life has shown him how to listen to the universe, to take notice of signs and warnings sent to him. When he did not do so, the only result was suffering. If that applies to one human being, why should that not apply to humanity?

Poet Elísabet Jökulsdóttir spoke to us through her living-room window, as she belongs to a group of triple vulnerability—due to her age, health, and mental condition. We fixed the camera, and she called us on the phone. She told us that there was nothing positive about this virus, she only felt terror, but it was up to us if we used this experience for the good or not. She gave us directions to a sculpture she had created in the village of Hveragerdi. It is a stone by a river, with a carved seat, where you can sit and watch the river pass by. The art piece is called *It will pass by* and it is dedicated to young people with mental problems and suicidal thoughts. Whatever you are dealing with, and however painful it seems, eventually it will pass by.

We went to visit theater director Thorleifur Arnarsson and his wife, visual artist Anna Rún Tryggvadóttir. They had taken the last flight to Iceland from Berlin after being in lockdown there for more than a month. Thorleifur had been rehearsing *Peer Gynt* in Vienna, but the show was canceled. "Theater is not only on hold," Thorleifur told us. "It died. Even during Nazi times, there was theater everywhere in Germany, official state theater supporting the regime and underground theater. But now there is nothing. It has died, it has died and online theater is not theater. Theater is all about live performance and the audience, and that is where the art takes place: in the connection of the two in flesh and blood. Now it is not possible; it's illegal and has been labeled a non-essential business. How can theater be the same again after this? And the same applies to our societies. We have just gone through one of humaniy's biggest social experiments. Are we just going to restart everything and pretend nothing happened?"

Thorleifur and Anna's eight-year-old son, Tryggvi, had not been in school for a full eight weeks. He had made a song about COVID-19 on the piano. He sat down, in his underwear, and started playing. Only playing the darkest keys to the far left, it sounded like an anthem, a frustrated scream of a generation of one hundred million children in lockdown. The lyrics went something like this: *Shit shit shit shit, ass ass ass ass ass I HATE COVID-19.*

We drove around the empty streets of Reykjavik to meet the artists and listened to the news: 2,000 people, almost 1 percent of Iceland's population, had lost their jobs in that single day. Most of them were connected to the airport and Icelandair. Tourism had become a significant section of our economy in recent years, bigger than the fisheries or any other industry. We had experienced a boom ever since the Eyjafjallajökull eruption put Iceland on the map as a tourist destination. Now it was totally wiped out, nobody knowing when travel would be permitted again.

Still, we were grateful for the actions taken here while news of chaos and death came from from Spain, Italy, the UK, and the U.S., accompanied by ever-weirder remarks from the American president.

To our surprise, the Icelandic government had shown professionalism and competence in managing the COVID-19 issue. All its decisions were science-based, and the orders did not come through politicians, but from daily meetings with our Chief Epidemiologist, the Director of Health, and the Chief Superintendent of the Icelandic Police. A likeable and humanistic group, they gave strict orders but emphasised individual responsibility.

Polls in Iceland measured 96 percent trust in the government's actions. We experienced anxiety and financial loss, but at the same time—and maybe for the first time since the 1990s, there was confidence and trust in the system and the government, the kind of trust you would give to a surgeon just about to put you to sleep and perform open-heart surgery. While uncritical belief in authority might not be healthy for the long term, after years of political polarization, this situation, this moment of trust in authority, was a strange relief. There was not a total lockdown, just strict two-meter distancing rules and a cap on how many people could enter a shop. Children could still go to school for up to two hours a day, as they did not seem to carry or catch the disease to the same degree as grown-ups. That idea proved to be a correct one, and it helped to keep many families and children sane through the crisis.

And so Iceland's government, with the help of Icelanders, managed to flatten the curve. There was extensive testing of almost 20 percent of the population. For those who were confirmed positive, there was precise tracing of those who they had been in contact with; all of them had to stay home in a fourteen-day quarantine. Almost 5 percent of the population—19,000 people—went through that. We all downloaded an app to help trace who we met and where we had been.

As I write this—on May 10, 2020—Iceland has had five days without a new COVID-19 case, down from 100 per day at its peak, and my children are hugging their grandparents again for the first time in two months. Our intensive care units never ran beyond capacity, and the survival rate of those patients was exceptionally high. Everyone who tested positive got a daily phone call from the hospital and if their symptoms got worse, they were called in. Of almost 2,000 cases, we had ten deaths.

Icelanders feel grateful for the lives that were saved in this way, and our low death rates by comparison with neighboring countries. Because most of the suffering was avoided, the common experience for most of the population was of being stuck at home, not meeting the older people in your life, having all the kids around, lost opportunities, and an unpredictable future of mass unemployment and fear of a second wave of the virus. That and the uncanny weirdness of everything.

Our interviews gave insights into a broad scope of thoughts. Some said the crisis would not change us, but rather it will make us even more hungry for the old ways, and that we will go running hungrier than ever into a new decade, a "Roaring Twenties" of consumption, production, speed, and waste. Others said that it proved governments could act according to science to prevent harm; that capitalism and the industrial machine was not a law of nature, above and unattached to science and human lives.

It is strange, but sometimes it feels like Icelanders feel more in their element during crisis times than during boom years. After a thousand years of living on one of the harshest places on earth, something is always going to happen eventually. Old women shake their heads on a good sunny day and mumble: "This is a bad omen." And something always does happen, of course. A frost winter, an epidemic, an avalanche, a volcanic eruption, a massive storm, or an earthquake. So now a small virus has kicked us off track. We have to rebuild and rethink. In a strange way, we knew this time would come. We had been joking about the tourist boom. "How will we use this hotel when the bubble bursts? An old people's home? An artists' residency?"

The apocalypse has happened across the globe. And it has been an unveiling of everything. We can see the structures: we can see the Himalayas, the sky over China is blue, the waters in Venice are clear. The virus has revealed, too, the imagination—or lack of it—of politicians and governments. When the chaos raged in Wuhan, many could not imagine that same situation in their own country and take the needed preventative steps. When it raged in Italy, it was still a faraway problem for northern Europeans and the U.S. The crisis has shown us the importance of understanding science and applying it to future realities. Yet we are slow to learn. Again and again, tragedy spreads because we don't believe it will happen to us. But—this time—everything was stopped, because a loved one might get sick next week. This danger spoke to the heart.

The big question now is: how can we act with the same urgency to protect the foundations of life for our loved ones in 2050, 2060, or 2080? Can we translate how global inaction caused immense suffering during the COVID crisis and apply that to the future of the whole planet? Was there anything in this great pause that can show us the way?

This book is dedicated to my children,
grandchildren, and great-grandchildren.

Andri Snær Magnason

Endnotes

1 "Corporate Default and Recovery Rates, 1920–2008," *Moody's Global Credit Policy*, www.moodys.com/sites/products/ DefaultResearch/2007400000578875.pdf.

2 "Splendor in the Mud: Unraveling the Lives of Anacondas," *The New York Times*, www.nytimes.com/1996/04/02/science/ splendor-in-the-mud-unraveling-the-lives-of-anacondas.html; Paul P. Calle, Jesús Rivas, María Muñoz, John Thorbjarnarson, Ellen S. Dierenfeld, William Holmstrom, W. Emmett Braselton, and William B. Karesh, "Health Assessment of Free-Ranging Anacondas (*Eunectes murinus*) in Venezuela," *Journal of Zoo and Wildlife Medicine*. Vol. 25, No. 1, Reptile and Amphibian Issue (March 1994), pp. 53–62.

3 *Vísir*, 10. March 1933, p. 2. In *Heimdalli*, 18. Apríl 1933 (p. 1) 507/5000, a chart of marine casualties that occurred in the first quarter of that year. In total, thirty-four people were killed; in addition, the article says that it is likely two German trawlers and one English also went down around Iceland, about thirty-six to forty people in total. Certainly fifty to one hundred Icelandic children were left fatherless during these three months. Icelanders lost proportionally as many at sea during these years as other nations did at war. 2014 was the first year in Icelandic history where no sailor died doing his job.

4 Helgi Valtýsson, *Á hreindýraslódum* [*In Reindeer Country*]. Akureyri 1945, p. 11.

5 Ibid, pp. 104–105.

6 "Our history," Alcoa, http://www.alcoa.com/global/en/who-we-are/history/default.asp.

7 Helgi Valtýsson, *Á hreindýraslódum* [*In Reindeer Country*], p. 57.

8 Sea Level Rise, http://atlas-for-the-end-of-the-world.com/ world_maps/world_maps_sea_level_rise.html.

9 Global Warming of 1.5°C. IPCC 2018, https://report.ipcc.ch/ sr15.

10 Anna Agnarsdóttir, "Var gerd bylting á Íslandi sumarid 1809?" ["Was there a revolution in Iceland in the summer of 1809?], *Saga XXXVII*, 1999.

11 *Íslenzk sagnablöd*, Kaupmannahöfn, 1826, p. 80. Translation: "Historical Account of a Revolution on the Island of Iceland in the Year 1809," ed. Ódin Melsted, Anna Agnarsdóttir (Reykjavík: Reykjavik University Press, 2016), p. 163.

12 K. Caldeira and M. E. Wickett, "Anthropogenic Carbon and Ocean pH," *Nature*, 2003, p. 425.

13 "Áhrif loftslagsbreytinga á sjávarvistkerfi" ["The impact of climate change on marine ecosystems."] *Morgunbladid*, 12. September, 2006, p. 10.

14 Cf. http://www.ipcc.ch/.

15 Heinrich Harrer. *Seven Years in Tibet* (London: Rupert Davis, 1953), p. 124.

16 *Dhammapada: vegur sannleikans. Ordskvidir Búdda.* [Dhammapada: The Way of Truth. Buddha's Sayings.] Trans. Njördur P. Njardvík (Reykjavík, 2003), p. 85.

17 *Landnámabók.* Íslensk fornrit I (Reykjavík: Jakob Benediktsson Publisher, 1986), p. 321.

18 P. Wester, A. Mishra, A. Mukherji, A. B. Shrestha (Eds.). The Hindu Kush Himalaya Assessment: Mountains, Climate Change, Sustainability and People. Springer Open, 2019. https://link.springer.com/content/pdf/10.1007%2F978-3-319-92288-1.pdf.

19 Ibid.

20 Kunda Dixit, "Terrifying assessment of a Himalayan melting," *Nepali Times*, February 4, 2019. www.nepalitimes.com/banner/a-terrifying-assessment-of-himalayan-melting/.

21 David Wallace-Wells, "UN Says Climate Genocide Is Coming. It's Actually Worse Than That," *Intelligencer*, http://nymag.com/intelligencer/2018/10/un-says-climate-genocide-coming-but-its-worse-than-that.html.

22 "Only 11 Years Left to Prevent Irreversible Damage from Climate Change, Speakers Warn during General Assembly High-Level Meeting," www.un.org/press/en/2019/ga12131.doc.htm.

23 Sea Level Rise Viewer, https://coast.noaa.gov/slr/.

24 Björn Thorbjarnarson: "The Shah's Spleen," www.journalacs. org/article/S1072-7515(11)00292-4/fulltext?fbclid=IwAR3u0_ tMdcN-ukZqwHhR7CfhNXrCPXkG4cz0ZBBKK8m zZY_LgpGA-hGIzVFg.

25 Blake Gopnik, "Andy Warhol's Death: Not So Simple, After All," *The New York Times*, www.nytimes.com/2017/02/21/arts/ design/andy-warhols-death-not-so-routine-after-all.html.

26 *The Decision to Drop the Bomb* (1965) dir. Fred Freed, Len Giovannitti

27 Wayne F. King, Harry Messel, James Perran Ross, and John Thorbjarnarson: "Crocodiles—An action plan for their conservation," https://portals.iucn.org/library/node/6002.

28 "Mamirauá, land flædiskógarins," *Morgunbladid*, August 24, 1997, pp. B18–19.

29 "Crocodiles—An action plan for their conservation," https:// portals.iucn.org/library/node/6002.

30 Lao Tse, Bókin um veginn. Jakob Jóhann Smári and Yngvi Jóhannesson thýddu (Reykjavík, 1921), p. 11.

31 Arctic Report Card: Update for 2018, https://arctic.noaa.gov/ Report-Card/Report-Card-2018.

32 Caspar A. Hallmann , Martin Sorg, Eelke Jongejans, Henk Siepel, Nick Hofland, Heinz Schwan, Werner Stenmans, Andreas Müller, Hubert Sumser, Thomas Hörren, Dave Goulson, Hans de Kroon: "More than 75 percent decline over 27 years in total flying insect biomass in protected areas," https://journals.plos.org/plosone/article?id=10.1371/journal. pone.0185809.

33 Douglas Martin, "John Thorbjarnarson, a Crocodile and Alligator Expert, Is Dead at 52," *The New York Times*, www. nytimes.com/2010/03/10/science/10thorbjarnarson.html.

34 "John Thorbjarnarson," *The Economist*, www.economist.com/ obituary/2010/03/18/john-thorbjarnarson.

35 P. Nielsen: "Sídustu geirfuglarnir." *Vísir*, 12. September 1929, p. 5.

36 P. Nielsen: "Sæörn." *Morgunbladid*, 2. August 1919, p. 2–3.

37 "Media Release: Nature's Dangerous Decline 'Unprecedented';
    Species Extinction Rates 'Accelerating'," IPBES—Science and
    Policy for People and Nature, www.ipbes.net/news/Media-
    Release-Global-Assessment.

38 "Nicholas Clinch, Who Took On Unclimbed Mountains, Dies
    at 85," *The New York Times*, www.nytimes.com/2016/06/23/
    sports/nicholas-clinch-who-took-on-unclimbed-mountains-
    dies-at-85.html

39 Sigurdur Thórarinsson, "Vatnajökulsleidangur 1956," *Jökull*.
    Ársrit Jöklarannsóknarfélags Íslands, p. 44.

40 María Jóna Helgadóttir, Breytileg stærd jökulsins Oks í
    sambandi vid sumarhitastig á Íslandi. [The Changing Size of
    the Ok glacier in relation to Icelandic summer temperatures.]
    BS dissertation in Earth Sciences at the University of Iceland,
    School of Engineering and Natural Sciences. Supervisor:
    Hreggvidur Norddahl. Háskóli Íslands, 2017.

41 Cf. www.facebook.com/jardvis/posts/2705880609426386.

42 H. Frey, H. Machguth, M. Huss, C. Huggel, S. Bajracharya,
    T. Bolch, A. Kulkarni, A. Linsbauer, N. Salzmann, and M.
    Stoffel, "Estimating the volume of glaciers in the Himalayan–
    Karakoram region using different methods," *The Cryosphere*,
    www.the-cryosphere.net/8/2313/2014/tc-8-2313-2014.pdf.

43 "CO_2 Concentrations Hit Highest Levels in 3 Million Years,"
    Yale Environment 360. The Yale School of Forestry and
    Environmental Studies, https://e360.yale.edu/digest/CO_2-
    concentrations-hit-highest-levels-in-3-million-years.

44 "Are Volcanoes or Humans Harder on the Atmosphere?,"
    Scientific American, www.scientificamerican.com/article/
    earthtalks-volcanoes-or-humans/.

45 "Planes or Volcano?," https://informationisbeautiful.net/2010/
    planes-or-volcano.

46 Calculations based on the assumption that the Eyjafjallajökull
    emissions were 150,000 tons of CO_2 per day, U.S. emissions
    amount to 5.4 gt and UK emissions to 400mt. Then methane
    and land use CO_2 equivalents can be added. The calculations
    here focus on just the fire alone.

47 Global Warming of 1.5°C. IPCC 2018, www.ipcc.ch/sr15. 46.
    See: www.globalcarbonproject.org/.

48 Global Warming of 1.5°C. IPCC 2018, https://report.ipcc.ch/sr15.

49 Hans Rosling, Ola Rosling, and Anna Rosling Rönnlund, "Deaths in Wars." Factfulness: Ten Reasons We're Wrong About The World—And Why Things Are Better than You Think.

50 Cf. www.hagstofa.is/utgafur/frettasafn/umhverfi/losun-koltvisyrings-a-einstakling.

51 "Extreme Carbon Inequality," Oxfam International, https://www-cdn.oxfam.org/s3fs-public/file_attachments/mb-extreme-carbon-inequality-021215-en.pdf.

52 Cf. http://www.loftslag.is/?p=10716.

53 For a good discussion of the theoretical basis of such calculations, cf. www.carbonbrief.org/analysis-how-much-carbon-budget-is-left-to-limit-global-warming-to-1-5c.

54 "Global energy transformation: A roadmap to 2050," IRENA—International Renewable Energy Agency, www.irena.org/publications/2019/Apr/Global-energy-transformation-A-roadmap-to-2050-2019Edition.

55 Cf. www.herkulesprojekt.de/en/is-there-a-master-plan/the-moon-landing.html.

56 "Mike Pompeo praises the effects of climate change on Arctic ice for creating new trade routes," *The Independent*, www.independent.co.uk/news/world/americas/us-politics/mike-pompeo-arctic-climate-change-ice-melt-trade-a8902206.html.

57 Sandra Laville, "Top oil firms spending millions lobbying to block climate change policies, says report," *The Guardian*, www.theguardian.com/business/2019/mar/22/top-oil-firms-spending-millions-lobbying-to-block-climate-change-policies-says-report.

58 "Annual Energy Outlook 2019," EIA—U.S. Energy Information Administration, www.eia.gov/outlooks/aeo/.

59 James Rainey: "The Trump administration scrubs climate change info from websites. These two have survived," NBC News, www.nbcnews.com/news/us-news/two-government-websites-climate-change-survive-trump-era-n891806.

60 "A Fifth of China's Homes Are Empty. That's 50 Million Apartments," Bloomberg News, www.bloomberg.com/news/articles/2018-11-08/a-fifth-of-china-s-homes-are-empty-that-s-50-million-apartments.

61 Cf. http://anthropocene.info/great-acceleration.php.

62 *Earth System Science Data*, www.earth-syst-sci-data-discuss.net/5/1107/2012/essdd-5-1107-2012.pdf.

63 "Revisiting the Earth's sea-level and energy budgets from 1961 to 2008," *Geophysical Research Letters*, https://agupubs.onlinelibrary.wiley.com/doi/full/10.1029/2011GL048794.

64 www.aims.gov.au/docs/research/climate-change/coral-bleaching/bleaching-events.html.

65 "Global warming impairs stock–recruitment dynamics of corals," *Nature*, www.nature.com/articles/s41586-019-1081-y.

66 Morgan Pratchett, "Coral reproduction on the Great Barrier Reef falls 89% after repeated bleaching." *The Conversation*. April 3 2019. https://theconversation.com/coral-reproduction-on-the-great-barrier-reef-falls-89-after-repeated-bleaching-114761.

67 *Loftslagsbreytingar og áhrif theirra á Íslandi.* [Climate change and its effect on Iceland.] Report of the Science Committee, 2018, "Ocean Acidification" (chapter 6), www.vedur.is/loftslag/loftslags-breytingar/loftslagsskyrsla-2018.

68 Jón Ólafsson, "Rate of Iceland Sea acidification from time series measurements," www.researchgate.net/publication/26636970.

69 Einar Jónsson, "Enn um átu," *Sjómannabladid Víkingur* 11–12, 1980, pp. 45–49.

70 Lee R. Kump, Timothy J. Bralower, and Andy Ridgwell, "Ocean Acidification in Deep Time," *Oceanography*, 4/22, 2009.

71 https://unfccc.int/sites/default/files/resource/The%20People%27s%20Address%202.11.18_FINAL.pdf.

72 J. E. N. Vernon et al., "The coral reef crisis: The critical importance of <350 ppm CO_2," *Marine Pollution Bulletin*, www.sciencedirect.com/science/article/pii/S0025326X09003816.

73 Ganopolski, A., Winkelmann, R. & Schellnhuber, H. Critical insolation-CO_2 relation for diagnosing past and future glacial inception. *Nature* 534, S19–S20 (2016). https://doi.org/10.1038/nature18452.

74 *Living Planet Report—2018: Aiming Higher.* WWF 2018, www.
wwf.org.uk/sites/default/files/2018-10/wwfintl_livingplanet_
full.pdf.

75 Gorman, James. "It Could Be the Age of the Chicken,
Geologically." *The New York Times.* Dec. 11, 2018. www.
nytimes.com/2018/12/11/science/chicken-anthropocene-
archaeology.html.

76 Christopher A. Brochu and Glenn W. Storrs, "A giant
crocodile from the Plio-Pleistocene of Kenya, the phylogenetic
relationships of Neogene African crocodylines, and the
antiquity of Crocodylus in Africa," *Journal of Vertebrate
Paleontology,* 32:3 (2012), pp. 587–602.

77 "Doha infographic gets the numbers wrong, underestimates
human emissions," *Carbon Brief,* www.carbonbrief.org/
doha-infographic-gets-the-numbers-wrong-underestimates-
human-emissions.

78 Cf. www.drawdown.org.

79 "U.S. could feed 800 million people with grain that livestock
eat, Cornell ecologist advises animal scientists," *Cornell
Chronicle,* https://news.cornell.edu/stories/1997/08/us-could-
feed-800-million-people-grain-livestock-eat.

80 Emily S. Cassidy, Paul C. West, James S. Gerber, and Jonathan
A. Foley, "Redefining agricultural yields: from tons to people
nourished per hectare," *Environmental Research Letters,* https://
iopscience.iop.org/article/10.1088/1748-9326/8/3/034015.

81 "National Inventory Report. Emissions of Greenhouse
Gases in Iceland from 1990 to 2017," Umhverfisstofnun
2019, *Umhverfisstofnun,* www.ust.is/library/Skrar/Atvinnulif/
Loftslagsbreytingar/NIR%202019%20Iceland%2015%20
April%20final_submitted%20to%20UNFCCC.pdf.

Photo credits

pp. 34–35 A sprint finish at Melavöllur Stadium. Unknown date and time. *Árni Kjartansson.*

pp. 36–37 Shovelling out the United States Army Skytrain plane at Bárdarbunga in 1951. *Árni Kjartansson.*

pp. 38–39 Hulda and her friend Hrefna rode to Stykkishólmur in the summer of 1939, aged almost fifteen. *Photograph by Hrefna Einarsdóttir. From the private collection of Hulda Gudrún Filippusdóttir.*

pp. 40–41 Hulda, aged sixteen or seventeen, on a Schulgleiter glider at the Icelandic Gliding Club in the summer of 1940 or 1941. *Photograph by Helgi or Pétur Filippusson. From the private collection of Hulda Gudrún Filippusdóttir.*

pp. 42–43 The twin sisters: Kristín (my mother) and Gudrún Björnsdóttir in front of a glider at the Icelandic Gliding Club, probably in the summer of 1950. *From the private collection of Hulda Gudrún Filippusdóttir.*

pp. 88–89 Kamadhenu and Audhumla. Audhumla image from the Edda manuscript written by the Reverend Ólafur Brynjólfsson in Hróarstunga in 1760, now in the holdings of the Danish Royal Library, Copenhagen, NKS 1867 4to. Kamadhenu image retrieved from a digital photo archive.

pp. 128–129 Jesus Christ and J. Robert Oppenheimer. The image of Jesus is an altarpiece from Bíldudalur Church (originally from Otradalur Church). The image is Danish and dates back to the year 1737. *Photograph by Högni Egilsson.* The image of Oppenheimer is by *Alfred Eisenstaedt/The Life Picture collection via Getty Images.*

pp. 150–151 Björn Thorbjarnarson (my maternal grandfather) and Arndís Thorbjarnardóttir. Photographs taken at Bíldudalur, around 1937.

pp. 152–153 These photographs, taken at the Tolkien family home in Oxford, England, in 1930, show Arndís, J. R. R. and Edith Tolkien and the four children: Christopher, John Francis Reuel, Priscilla, and Michael. *From the private collection of Arndís Thorbjarnardóttir. Photographers: Arndís Thorbjarnardóttir and J. R. R. Tolkien.*

pp. 158–159 Árni Kjartansson and Hulda Gudrún Filippusdóttir at Hvannadalshnúkur in May 1956. *Photograph taken by their friend, Ingibjörg Árnadóttir.*

pp. 160–161 On the way to Kverkfjöll, May 1954. In the photograph are, from front to back: Haukur Haflidason, Steina Audunsdóttir, Sigurdur Waage, probably Stefán Bjarnason, Hulda Gudrún Filippusdóttir, and Árni Kjartansson. *Árni Kjartansson.*

pp. 162–163 Making camp during the honeymoon journey in 1956. *Árni Kjartansson.*

pp. 164–165 The group, on skis, being pulled across Vatnajökull, 1956. *Árni Kjartansson.*

pp. 196–197 *Activ of London* and an oil or gas platform near Chacachacare Island in the Republic of Trinidad and Tobago, 2018. *Martin de Thurah.*

pp. 232–233 *Activ of London*, 2018. *Martin de Thurah.*

pp. 234–235 Sea turtle in the Tobago Cays, Tobago. *Martin de Thurah.*

pp. 236–237 Pelicans at Englishman Bay, Tobago. *Martin de Thurah.*

pp. 264–265 A sacred cow in Dharamshala, India, June 2010. *Arnar Thórisson.*

pp. 266–267 Tenzin Gyatso, the fourteenth Dalai Lama. A still from a recorded interview in Dharamshala on June 9, 2010. *The videographers were Dagur Kári Pétursson and Arnar Thórisson.*

pp. 288–289 John Thorbjarnarson. The first photograph is taken from a family album and shows him aged about ten years old. The second is from 2007 with a Chinese alligator.

pp. 290–291 *Crocodylus thorbjarnarsoni.* A scientific diagram based on fossil records. *Image by Christopher Brochu and Glenn Storrs.*

pp. 310–311 Kristín and Gudrún Björnsdóttir's tenth birthday at number 3 Selás, 1956. A still from a 16 mm film. *Árni Kjartansson.*

Index

Note: Most Icelandic names comprise a given name plus a patronymic (ending –son or –dóttir) that indicates family origin but usually changes with each generation. Inverting these seems unhelpful so, while Charles Lindbergh appears at "L," John Thorbjarnarson will be found under "J" rather than "T." *Italic* page references lead to full-page illustrations; "n" refers to one of the numbered endnotes (indexed only very selectively) after page 324.

India
    author's first impressions 261–2
    Dharamsala 81, 108, 261–2,
        264–5, 269–70
    inequality 285–7
    secularism and religion 277
Indus, River 86, 111
infant mortality 30, 52, 72
Ingibjörg Árnadóttir 158–9, 168
insects 137, 327n32
"The Institute" see Árni
    Magnússon Institute for Medieval
    Studies
"interesting times" curse 10
international cooperation prospects
    182, 301
IPCC (UN Intergovernmental
    Panel on Climate Change) 62, 78,
    115, 195, 210
Iran, U.S. embassy seizure (1979)
    122
Israel 215–16

J

Jesus Christ 125, 128
job losses 319, 321
John Thorbjarnarson (author's
    uncle; son of Björn and Peggy)
    crocodile conservationist 130–1,
        138, 288, 289, 292–4, 327n27–9
    South American expedition 12,
        226
Jökulheimar research base 45–6,
    166–8, 170, 175
Jökull (journal) 49, 157, 168, 170
Jón Eythórsson 167
Jón Ólafsson 238
Jón Pétursson see Grandpa Jón
Jón Samsonarson 258
Jón Sigurdsson 74
Jónas Hallgrímsson 74
Jørgen Jørgensen ("King of the Dog
    Days") 70–5, 77–8, 242, 259

K

Kailas, Mount (Tibet) 85, 92, 98–9,
    284
Kamadhenu see cows; Hindu
    mythology
Kaupthing Bank 7–8
Keeling, Charles 174, 238
Ketill Ketilsson 140, 143–4
the "King's Book" 13–16
Kjartan Vigfússon (Grandpa Árni's
    father) 26–7
Kringilsárrani district
    Helgi Valtýsson's enthusiasm for
        49, 51–2, 55, 60
    inspiration for Dreamland 63
    reservoir scheme 54, 57–8, 59–60,
        213, 217
Kristín Björnsdóttir (author's
    mother) 42–3, 145, 310–11
    as an identical twin with Gudrún
        25, 44
Kvaran, Gunnar 317

L

landscapes, as ephemeral 251, 254
language
    of the climate emergency 78, 207
    dependence on culture 68–9,
        73–5, 77, 132
    evolving alphabets 91
    of freedom and independence
        70–3, 242, 259
    of Hallgrímur Pétursson 68–70
    of Helgi Valtýsson 50–2
    Icelandic and Faeroese compared
        70
    Icelandic and Hindi compared 87
    inadequacy in times of crisis 8–9,
        61–2
    Indo-European group 90
    of Jørgen Jørgensen 70–1
    of moderation and liberalism 56
Laozi, on usefulness 136
Laxness, Halldór 49, 303

T

technology
  and conservation 186–7
  downsides 297
  and overconfidence 280–1
  temperature effects on coral 231–8
  temperature targets *see* global
    warming
temperatures, global 61, 107, 155,
  198
tern, arctic 222, 249, 253, 258
tern, black 228
"terraforming" the Earth 300
theater, death of 318–19
Thompson, Lonnie 179–82
Thórhallur Filippusson (Grandma
  Hulda's brother) 30
Thorleifur Arnarsson 318–19
Thunberg, Greta 298
Tibet
  Chinese invasion and aftermath
    80, 83, 95–7, 105–5, 211, 263
  glacier melting 180–1
  Mount Kailas 85, 92, 98–9, 284
  Tibetan community in India
    269–70
  Tibetan Plateau as a third pole 98
tidal deposits 222–3
timescales
  family memories 18, 21–2
  folksong tradition 17
  lifetime of documents 125
  limitations of predictions 117, 295
  measuring by generations 241–2,
    308–9
  of tragedies and responses 322
tipping points 243, 246, 298
Tobago and Tobago Cays 196–7,
  225–6, 230, 234–7, 246
Tolkien family 148–9, 152–3
tourism 56, 136–7, 316, 319, 322
Trampe, Count Frederich 71
transport sector emissions 200,
  205, 299–301
Trausti, Jón 200

trawler, *Arinbjörn Hersir* 25–6
Trump Administration 209–10, 319
Tryggvi (son of Thorleifur and
  Anna) 319
turtles 130–1, 185, 246
  Chelonia mydas 228, *234–5*
twins, identical 25, 44

U

Uganda 138, 292, 294
unemployment 319, 321
United Nations
  climate goals 107
  Framework Convention on
    Climate Change Conference
    (2018) 244
  Intergovernmental Panel on
    Climate Change 62, 78, 115,
    195, 210
  Intergovernmental Science-Policy
    Platform on Biodiversity and
    Ecosystem Service 142
  Sustainable Development Goals
    301–4
United States
  climate change denial 209–10
  per capita emissions 201
Unnur Elísabet (Gunnarsdóttir)
  316–17

V

Valhalla 15, 84
Valley of Dreams (*Draumadalur*)
  46, 307
Valur (Filippusson, Grandma
  Hulda's brother) 30–1, 219
Vatnajökull glacier
  aircraft wreck on 23–4, 171
  Iceland Glaciological Society 1955
    expedition 45–6
  Iceland Glaciological Society
    1956 expedition 156–7, *158–65*,
    166–71
  July 2012 expedition 183–5

Inga Ābele (Latvia)
High Tide
Naja Marie Aidt (Denmark)
Rock, Paper, Scissors
Esther Allen et al. (ed.) (World)
The Man Between: Michael Henry Heim & a Life in Translation
Bae Suah (South Korea)
A Greater Music
North Station
Zsófia Bán (Hungarian)
Night School
Svetislav Basara (Serbia)
The Cyclist Conspiracy
Michal Ben-Naftali (Israel)
The Teacher
Guðbergur Bergsson (Iceland)
Tómas Jónsson, Bestseller
Max Besora (Spain)
The Adventures and Misadventures of Joan Orpí . . .
Jean-Marie Blas de Roblès (World)
Island of Point Nemo
Per Aage Brandt (Denmark)
If I Were a Suicide Bomber
Can Xue (China)
Frontier
Vertical Motion
Lúcio Cardoso (Brazil)
Chronicle of the Murdered House
Sergio Chejfec (Argentina)
The Dark
The Incompletes
My Two Worlds
The Planets
Eduardo Chirinos (Peru)
The Smoke of Distant Fires
Manuela Draeger (France)
Eleven Sooty Dreams
Marguerite Duras (France)
Abahn Sabana David
L'Amour
The Sailor from Gibraltar
Mathias Énard (France)
Street of Thieves
Zone
Macedonio Fernández (Argentina)
The Museum of Eterna's Novel
Rubem Fonseca (Brazil)
The Taker & Other Stories
Rodrigo Fresán (Argentina)
The Bottom of the Sky
The Dreamed Part
The Invented Part
Juan Gelman (Argentina)
Dark Times Filled with Light

Oliverio Girondo (Argentina)
Decals
Georgi Gospodinov (Bulgaria)
The Physics of Sorrow
Arnon Grunberg (Netherlands)
Tirza
Hubert Haddad (France)
Rochester Knockings: A Novel of the Fox Sisters
Gail Hareven (Israel)
Lies, First Person
Angel Igov (Bulgaria)
A Short Tale of Shame
Ilya Ilf & Evgeny Petrov (Russia)
The Golden Calf
Zachary Karabashliev (Bulgaria)
18% Gray
Ha Seong-nan (South Korea)
Bluebeard's First Wife
Flowers of Mold
Hristo Karastoyanov (Bulgaria)
The Same Night Awaits Us All
Jan Kjærstad (Norway)
The Conqueror
The Discoverer
Josefine Klougart (Denmark)
One of Us Is Sleeping
Carlos Labbé (Chile)
Loquela
Navidad & Matanza
Spiritual Choreographies
Jakov Lind (Austria)
Ergo
Landscape in Concrete
Andri Snær Magnason (Iceland)
On Time and Water
Andreas Maier (Germany)
Klausen
Lucio Mariani (Italy)
Traces of Time
Sara Mesa (Spain)
Four by Four
Amanda Michalopoulou (Greece)
Why I Killed My Best Friend
Valerie Miles (World)
A Thousand Forests in One Acorn
Subimal Misra (India)
This Could Have Become Ramayan Chamar's Tale
Iben Mondrup (Denmark)
Justine
Quim Monzó (Catalonia)
Gasoline
Guadalajara
A Thousand Morons
Why, Why, Why?

OPEN LETTER

WWW.OPENLETTERBOOKS.ORG